AN ANCIENT
MESOPOTAMIAN
HERBAL

AN ANCIENT MESOPOTAMIAN HERBAL

BARBARA BÖCK, SHAHINA A. GHAZANFAR & MARK NESBITT

Kew Publishing
Royal Botanic Gardens, Kew

Royal Botanic Gardens Kew

MINISTERIO DE CIENCIA E INNOVACIÓN

UNIÓN EUROPEA
FONDO EUROPEO DE DESARROLLO REGIONAL
"Una manera de hacer Europa"

AGENCIA ESTATAL DE INVESTIGACIÓN

Project PGC2018-097821-B-I00 and PID2021-125678NB-I00 funded by MCIU/AEI/10.13039/50110, MICINAEI/10.13039/501100011033 and FEDER A way to make Europe

First published in 2023 by
Royal Botanic Gardens, Kew,
Richmond, Surrey, TW9 3AB, UK
www.kew.org

ISBN 978 1 84246 798 5
e-ISBN 978 1 84246 799 2

Distributed on behalf of the Royal Botanic Gardens, Kew in North America by the University of Chicago Press, 1427 East 60th St, Chicago, IL 60637, USA.

British Library Cataloguing in Publication Data
A catalogue record for this book is available from the British Library

Design and page layout: Nicola Thompson, Culver Design
Project management: Georgina Hills
Copy-editing: Matthew Seal
Proofreading: Sharon Whitehead

Printed and bound in Great Britain by Short Run Press

FSC
MIX
Paper from responsible sources
FSC® C014540
www.fsc.org

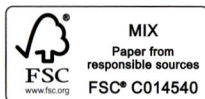

For information or to purchase all Kew titles please visit shop.kew.org/kewbooksonline or email publishing@kew.org

Kew's mission is to understand and protect plants and fungi, for the wellbeing of people and the future of all life on Earth.

Kew receives approximately one third of its funding from Government through the Department for Environment, Food and Rural Affairs (Defra). All other funding needed to support Kew's vital work comes from members, foundations, donors and commercial activities, including book sales.

CONTENTS

PREFACE

Throughout the world, traditional medical systems exist alongside conventional (or 'Western') medicine. These systems vary greatly in their underlying beliefs – but all have in common the consumption of medicine by the patient as a central practice. Most of these medicines are plant-based; even in today's conventional medicine, many pharmaceutical drugs have an origin in the natural world. The study of *materia medica* – medicinal drugs – can reveal much about a society: where did the drugs come from? Who prescribed them? Who took them? What were the beliefs (and the actual mechanism, if any) behind their function? How was knowledge codified and passed down the generations? What does use of medicines tell us about the complicated relationship between mind and body that is implicated in illness and recovery?

Ancient Mesopotamia makes a particularly rich – though at times tantalising – case study. Its cuneiform clay tablets date back over 5,000 years, forming the oldest written record of any medical system. We are fortunate in being able to draw on a detailed record of the natural resources of today's Mesopotamia – the country of Iraq, and neighbouring regions – which has been well documented by naturalists. Furthermore, natural products such as plants, animals and minerals are still used locally in traditional medicine, forming part of the region's rich cultural heritage. The medicines of ancient Mesopotamia are truly foundational: there is strong evidence for the transmission of their use to classical Greece, and thence both to today's Western herbalism, and eastwards to the Unani medical system of South Asia.

Almost 100 years have passed since the British Assyriologist, archaeologist and British Museum curator, Reginald Campbell Thompson (1876–1941), published *The Assyrian Herbal*, a paper he first presented in March 1924 at the Royal Society in London, reproduced directly from his handwritten manuscript. In Thompson's herbal, he identified about 250 drugs of vegetable origin; his posthumously published *A Dictionary of Assyrian Botany* (1949) is a revised edition of the 1924 version. Thompson was in many regards a pioneer, and these two books remain the only comprehensive studies of the ancient Mesopotamian pharmacopoeia. However, with the passing of time, new cuneiform material has appeared, outdating many of his readings and interpretations; the plants of current-day Iraq have been systematically studied and a new academic field, archaeobotany, has led to the identification of ancient seeds and charcoal from archaeological excavations.

Palm groves in the valley of the Euphrates, 250 miles northwest of Babylon, Iraq, c. 1915. From a stereoscopic photograph, Underwood & Underwood
Courtesy: Library of Congress (95517695)

The complex task of convincingly identifying ancient plant names can only be tackled with collaboration between the humanities and the sciences. This book is the result of such a dialogue, joining the efforts of Barbara Böck, Assyriologist specialising in ancient Mesopotamian medicine, Shahina Ghazanfar, botanist and co-editor of the *Flora of Iraq*, and Mark Nesbitt, archaeobotanist and curator of the Economic Botany Collection at the Royal Botanic Gardens, Kew. The book draws on all our skills. In Part 1, we introduce the lands and ancient cultures of Mesopotamia, survey the different forms of evidence that we use, explain what is known about the use of herbal medicines of ancient Mesopotamia, and describe our methods in researching this book. In Part 2, we discuss 25 plant-based drugs, which are obtained from plants ranging from agricultural crops and culinary spices to trees and wild-grown herbs. Among the plant names and terms for plant parts identified for the first time are red bryony and black bryony, manna (see under tamarisk), tamarisk gallnut and cedar tar. Other existing identifications have been clarified, such as poplar, which was thought to be Euphrates poplar but is perhaps the white poplar.

For each plant entry, the general context of the plant according to cuneiform sources, the meaning of the Akkadian and Sumerian name(s), cognates or equivalent words in Arabic or Hebrew, the etymology of the Latin and English terms, and the translations of the Akkadian and Sumerian texts are by the first author, Barbara Böck. Shahina Ghazanfar helped with the identifications, and with the scientific names and their meanings. Shahina and Mark Nesbitt are responsible for botanical descriptions, plant distributions, and traditional uses in Iraq and the Near East, and Mark for archaeobotanical data. Uses according to Graeco-Roman works are by Barbara, and uses according to Islamic works by Barbara and Shahina. All of the authors revised and agreed the whole text.

This book is dedicated to the people of modern-day Iraq, the custodians of a remarkable natural and cultural heritage.

Madrid and Kew
BB, SAG, MN

A NOTE ON PLACE NAMES, DATING AND ORTHOGRAPHY

Places are usually referred to by their ancient names; if the name is not known or widely familiar, the name of the modern site is given in the form most common in the Assyriological and archaeological literature. As for the dates, we use the so-called Middle Chronology when citing individual Mesopotamian rulers. Other dates are based on calibrated radiocarbon dating. We use diacritical marks for the transliteration of Sumerian, Akkadian, Arabic, Syriac and Hebrew plant terms; when they are cited, they are italicised, as are Latin and Greek plant terms. With the exception of Arabic, personal and place names are transliterated without diacritical marks.

Vowels

a like /a/ in "had"
ā like /a/ in "father"
e like /e/ in "yet"
ē like /ea/ in "head"

i like /i/ in "hit"
ī like /ee/ like in "eel"
u like /u/ in "put"
ū like /oo/ in "goose"

Consonants

ʾ   like the vocal break heard in the expression "uh-oh"

ʿ   a harder guttural sound only heard in Arabic

b   like /b/ in "band"

d   like /d/ in "dance"

ḍ   like /d/ in "bulldozer"

ḏ   like /th/ in "there"

f   like /f/ in "fish"

g   like /g/ in "good"

ǧ   like Parisian French /r/

h   like /h/ in "house"

ḥ   like /h/ in "Hamburg"

ḫ   like /ch/ in "loch"

j   like /j/ in "jar"

k   like /k/ in "kite"

l   like /l/ in "love"

m   like /m/ in "moon"

n   like /n/ in "noon"

p   like /p/ in "pot"

q   similar to /q/ in "queen" but pronounced further back in the mouth

r   like Scottish rolling /r/

s   like /s/ in "see"

ṣ   like /ts/ in "tsunami"

š   like /sh/ in "shadow"

t   like /t/ in "tree"

ṭ   like /t/ in "tut-tut"

ṯ   like /th/ in "think"

w   like /w/ in "wine"

y   like /y/ in "yes"

z   like /z/ in "zoo"

ẓ   like /z/ in "Zorro"

PART 1

INTRODUCTION

1: THE LAND BETWEEN THE RIVERS

Mesopotamia, the 'land between the rivers' Euphrates and Tigris as the ancient Greeks termed it, was home to many different groups of people, each speaking a different language. Sumerians, Akkadians, Assyrians, Babylonians and Chaldeans dominated and lived in the land from at least the 4th millennium BCE to the end of the 1st century CE, the period of time covered by the cuneiform sources. As a historical region, Mesopotamia extended over present-day Iraq and parts of Syria, Turkey, Iran and Kuwait. It is marked by stark geographical contrasts: mountain ranges and forest in the north and northeast flanking the region of upper plains and foothills; vast desert areas in the south and southwest; and the great alluvial plain and marsh areas of lower Mesopotamia in the southeast (see map). The two principal rivers, the Euphrates and Tigris, run from northwest to southeast before flowing into the Persian Gulf, each river taking a different course of travel. Both rise in eastern Turkey in the Taurus Mountains within 80 km of each other. The middle Euphrates traverses northern Syria, the middle Tigris northern Iraq; both empty their lower courses into the alluvial plain of central Iraq. It is their middle and lower courses that have, most of all, enabled, shaped and determined the life of the ancient Mesopotamian people. In geographical terms, the middle courses and their adjacent areas comprise the region of Assyria, home of the Assyrians, the lower courses Babylonia, home of the Babylonians and before them the Sumerians.

Sea of Azov

Crimea

B l a c k   S e a

R. Drava

R. Sava

Transylvanian Alps

River Danube

Balkan Mts.

Rhodope Mts.

Dinaric Mts. 6,581

Istria

Adriatic Sea

MACEDONIA

Sea of Marmara

Bosphorus

Hellespont

A S I A   M I N O R

Mt. Ida 5,797

Lesbos

Mt. Olympus 9,573

Apennine Mts.

Rome

Vesuvius 3,891

G. of Taranto

Corfu (Corcyra)

Pindus Mts. 8,645

GREECE

Chios

Samos

Euboea

Athens

Cyclades

Ephesus

Taurus Mts.

L. Tuz 10,672

10,125

10,696

11,762

12,369

11,315

SYRIA

FERTILE

Antioch · Ebla

Ugarit

Ionian Sea

Cephalonia

Zacynthos

Sparta

2,905

Peloponnese

Cythera

Khania (Canea)

Rhodes

Carpathos

CRETE

CYPRUS

Nicosia

Byblos

10,131

8,267

Sidon

Tyre

Damascus

Syrian Desert

Ri

SICILY

Etna 10,741

Malta

M e d i t e r r a n e a n   S e a

Joppa

PALESTINE

Jerusalem

Dead Sea 1,285

W. Sirhan

Dumah

Alexandria

LOWER EGYPT

Nile Delta

Memphis

L. Qarun

3,576

Sinai Penin.

G. of Aqaba

CYRENAICA

Libyan Plateau

Qattara Depression -440

Siwa Oasis

UPPER

Gulf of Suez

Red Sea Hills

7,910

Tema

A

L I B Y A

Jalo Oasis

W e s t e r n

Farafra Oasis

EGYPT

R. Nile

Eastern Desert

Jebel Shāïb

Libyan Desert

D e s e r t

Thebes

R e d   S e a

Dakhla Oasis

Kharga Oasis

1st Cataract

6,352

ETHIOPIA

7,270

2,652

River Nile · Abu Simbel

Nubian Desert

9,410

Map of Mesopotamia – Overview

© Oxford University Press, reproduced with the permission of the Licensor through PLSclear, map produced by Oxford Cartographers 98800

Landscapes

Each region was and is characterised by a distinct climate and vegetation. Farming in the lower Mesopotamian region depended on artificial irrigation. The main crops were barley, emmer, sesame and flax. Onions, leek, garlic, various legumes, vegetables, herbs and spices, pomegranates and figs were cultivated in gardens and orchards, usually intercropped with the ubiquitous date palm. In the non-irrigated zones of the alluvial plain grew camel thorn and false carob, in desert areas colocynth, and in the wadis greater plantain. Rain-fed agriculture

was possible in parts of the upper plains and foothills but was supported by local irrigation systems. The main crops in the north included barley, wheat and sesame; vegetables, herbs and fruits were cultivated in gardens and orchards. In ancient times, the foothills were covered with forests of oak, ash and terebinth (*Pistacia terebinthus*). In the upper hill zones, wildflowers and herbs flourished for much of the year, but in the steppe, the growing season was short with plants appearing in spring and dying off a few months later in the scorching heat of May and June. Along the rivers grew reeds, poplar, willow, tamarisk and liquorice.

FIGURE No.15

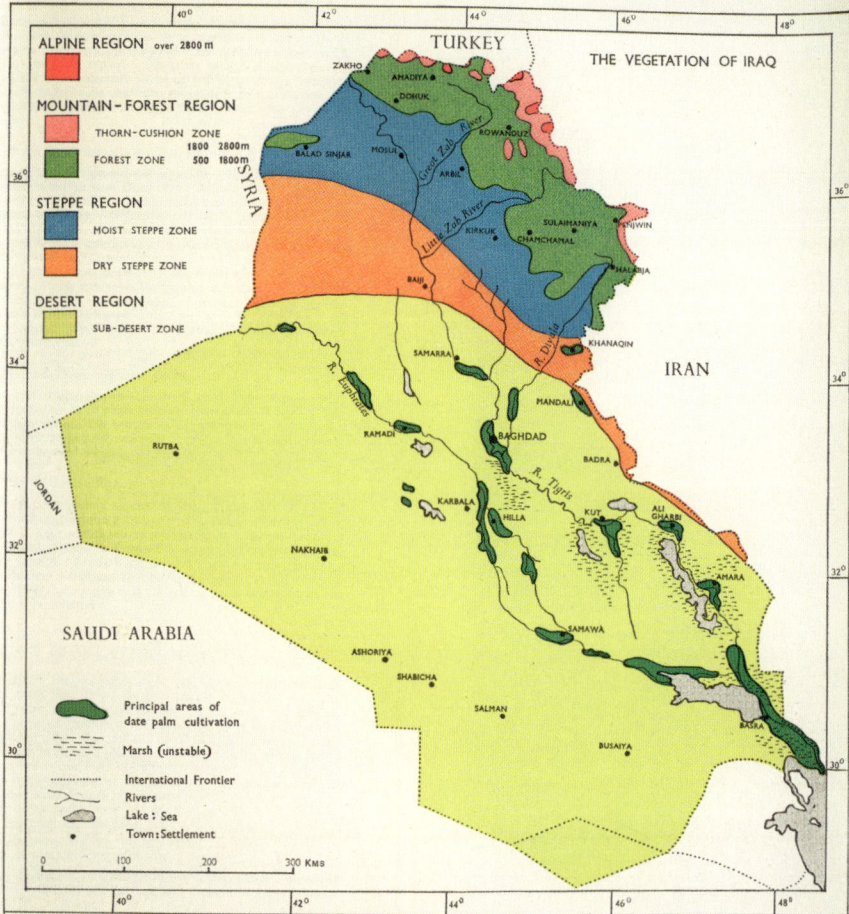

Physiographic regions and districts of Iraq
Reproduced with permission from *Flora of Iraq* 1: 63

Given the harsh environment, the creation of landscaped gardens by Assyrian and Babylonian kings was a natural choice. The earliest account of the creation of such a garden dates to the Assyrian king Tiglath-Pileser I (1115–1077 BCE) who planted opulent orchards and lush gardens close to his royal palace in Nineveh. Four centuries later, another Assyrian king, Sennacherib (704–681 BCE) prided himself on planting in the same city a 'replica of Mount Amanus', the Nur Mountains that rise from the Mediterranean Sea in southern Turkey. He imitated the wooded land and hilly terrain, famous for its cedars, and constructed aqueducts and hydraulic devices to ensure sufficient irrigation.

18

The huge park included a part devoted to Chaldea, in which the king recreated the typical landscape of southern Mesopotamia with its reeds and marshes. He was also reported to cultivate 'trees that bear wool' – cotton. Two thousand years ago, Greek and Latin authors credited the Babylonian king Nebuchadnezzar II (605–562 BCE) with the creation of one of the seven wonders of the ancient world, the Hanging Gardens of Babylon. If we believe the Roman Jewish historian Flavius Josephus (1st century CE), the king erected in the plain of Babylon massive hills of stones resembling natural mountains, and planted the trees of northwestern Iran to make his Median wife feel as if she were in her native country. However, no trace of this garden has been found at Babylon, and it is notably absent from written records before the classical period.

Languages

The peoples of ancient Mesopotamia spoke and wrote principally in Sumerian and Akkadian. Sumerian is an isolated language with no known parent; its first written records date to the late 4th millennium BCE. The language, which

Sculptured relief panel from Ashurbanipal's North Palace at Nineveh showing the royal landscape garden watered by an aqueduct, possibly built by Sennacherib, the king's grandfather, c. 645–635 BCE
© The British Museum, BM 124939,a

flourished during the 3rd millennium, is associated with the communities living in southern Mesopotamia. Sumerian was gradually substituted as a spoken language by Akkadian, which was written at least from the first half of the 3rd millennium. However, Sumerian lived on as a literary language and vanished eventually, together with Akkadian, with the last written cuneiform tablet in Akkadian dating to the 1st century CE. Akkadian is a Semitic language and is thus related to Arabic, Aramaic, Hebrew and Syriac. It was spoken in two dialects: Assyrian, used in northern Mesopotamia, and Babylonian, used in the south.

Writing

Sumerian and Akkadian were both written in cuneiform script, which was first developed for the Sumerian language. The script was named cuneiform (from Latin *cuneus* "wedge") by scholars in Europe from the characteristic wedge-shaped impressions that scribes impressed on the moist surface of clay tablets with a reed stylus. Cuneiform writing evolved in the context of bookkeeping and accounting, which required the storage of information about goods, quantities and containers. The earliest clay tablets found at the city of Uruk in the second half of the fourth millennium BCE record combinations of numerical notations and signs, representing plants, animals, textiles or metals and names. In order to learn the cuneiform signs and the language, the scribes working in the administration produced so-called lexical texts, thematic lists with words for titles and professions, god names and all types of commodities. Among these early lexical texts are also plant lists mentioning, for example, fig, grape and onion.

The earliest signs were pictographic, as with Egyptian hieroglyphs, in other words "barley" and "ox" were represented by a drawing of an ear of corn and the head of an ox, respectively. Over time, the drawings became simplified and more abstract.

Evolution of the sign for barley, Sumerian *še*

Evolution of the sign for ox, Sumerian *gu₄*

By 2700 BCE, these pictographs had evolved to the standard symbols that remained in widespread use for almost three millennia. However, an initial suite of 2,000 signs was reduced to 600 in about 1700 BCE. The number of cuneiform signs used depended on the text genre – thus letters and administrative records were written with fewer than 200 signs, while literary compositions used a far greater repertoire of signs. As in modern English words, signs performed different functions: as a syllable, as a whole word or an idea, or as a signifier, indicating what kind of word follows or precedes. For example, and relevant to this book, the signs u_2 "herb", *giš* "tree, bush" and *šim* "aromatic" were written before a plant name, classifying the nature of the plant, and the sign *šar* "green plant" was attached at the end of a plant term.

u_2 𒌑

giš   𒄑

šim   𒋆

šar   𒊬

First developed for the Sumerian language, cuneiform proved a highly adaptable script, which was used for Semitic Akkadian from 2500 BCE, and then for languages such as Indo-European Hittite, spoken in Anatolia, and Elamite in the Iranian plateau.

Clay has valuable properties as a writing material. It was widely available in the river valleys of Mesopotamia; no ink or special implement beyond the commonly available reed stylus was required and, once dried, a clay tablet could not be tampered with. Tablets are also portable, hence well-suited to correspondence over long distances. They are robust, even more so when baked in the ovens of Assyrian libraries, and they have little intrinsic value or potential for reuse, so remained relatively immune from casual destruction or looting until European collectors created a demand in the mid-nineteenth century. Tablets are however vulnerable to breakage and are often found as fragments; when a text is repeated on different tablets, it is sometimes possible to fill in the gaps in the original text.

Cuneiform tablets offer wonderfully direct access to the past. The clay tablet read by an Assyriologist is exactly that inscribed anywhere up to 4,000 years ago. This is a very different position from the study of ancient Greek and Roman medicine, which depends on texts copied into different formats and different languages during their transmission to today. The cuneiform textual material is large and diverse. About one million clay tablets have been excavated over the past 150 years and are housed in museums, of which about a quarter have been published. As discussed below and in Chapter 4, there is great unevenness in the representation of periods and regions, depending in part on which sites have been excavated (or looted), and in part on the nature of Mesopotamian society at a given time. In particular, the periods of imperial expansion discussed below were often accompanied by an emphasis on administration, and thus the creation of large archives of clay tablets.

Although the written evidence has major chronological gaps, in part linked to a long-term cycle of alternations of centralised government and turmoil, there is nonetheless a remarkable continuity in Mesopotamian civilisation over the period 3300–400 BCE.

The 3rd and 2nd millennia BCE

The political system of the Early Dynastic period (2900–2330 BCE) was marked by independent city-states that formed city-leagues. Important capitals in the south were Ur, Uruk, Adab, Kish, Umma, Lagash, Nippur and Sippar. Our entry on poplar (page 149) includes an image of the headdress of Puabi, who belonged to the 25th-century BCE ruling class of Ur. Her body was buried in the famous Royal Cemetery of Ur, which Sir Leonard Woolley excavated during the 1920s and early 1930s. We have used administrative texts belonging to this period. The Early Dynastic period also saw the creation of the oldest cuneiform herbal, dating to about 2450–2300 BCE. This tablet was found in the city of Ebla, located 55km southeast of Aleppo and is written in the Eblaite language, a Semitic language closely related to Akkadian. Ebla was one of the earliest kingdoms in Syria. At the peak of its power (about the middle of the 3rd millennium BCE) it dominated northern Syria, Lebanon and parts of northern Mesopotamia.

In the last century of the 3rd millennium BCE (2110–2003 BCE) the so-called Third Dynasty of Ur ruled over parts of Mesopotamia. The kings were based in the city of Ur, from where they exercised power over more than 20 cities located in the alluvial plain. The founder Ur-Namma (2110–2093 BCE) initiated a major building programme; temple towers, the famous ziggurats, were erected throughout the region under his rule. The political influence of the king extended further to the north, including cities located in the middle courses of the Euphrates and Tigris rivers. To control the state budget and distribute supplies among the cities, the king established effective organisational structures, with numerous state-run archives improved by the second king of the dynasty, Šulgi (2092–2045 BCE). The period is associated with the progressive disappearance of Sumerian as a vernacular language. Administrative texts from this period are a highly informative source for agricultural crops and garden products. Furthermore, the oldest surviving medical recipes, written in Sumerian, date to this period.

Population movements mark the beginning of the 2nd millennium BCE. Amorite dynasties, perhaps originating in Arabia, took over several important cities in the north and south of Mesopotamia, including Ashur, Isin, Larsa, Babylon, Eshnunna and Mari, the latter town being an important perfume centre where sweet flag was processed in great quantities. The most powerful of these dynasties was that of Babylon (1894–1595 BCE). Its first major ruler, Hammurabi (1792–1750 BCE), is best-known for his collection of laws. Four out of over 280 laws stipulate the cultivation and pollination of date palms within orchards. Hammurabi's reign extended over the cities located in the alluvial plain, including those situated along the middle course of the Euphrates as far as Mari. The spoken language was Akkadian in its Babylonian dialect. This period saw the production of most of the Sumerian literary texts known today, which are known from tablets produced as writing exercises in schools. Also written down during this time were the oldest Babylonian Akkadian medical prescriptions, a small group of culinary recipes, and literary compositions such as the Babylonian epic of the legendary king Gilgameš.

In cultural terms, the second half of the 2nd millennium BCE witnessed systematic recording of knowledge of Akkadian literature and science. The oldest copies of the glossary on medicinal ingredients, titled *Uruanna*, were produced around the 12th century BCE in Ashur. Situated on the western bank of the Tigris, the city-state marked the borderline between rain-fed and irrigation agriculture. It flourished from the beginning of the 2nd millennium BCE, when it had become an important trading platform, and became the capital of Assyria between the 14th to 9th centuries BCE. The spoken language was Akkadian in its Assyrian dialect. Perhaps it is no coincidence that the glossary was composed around the time when the oldest known landscaped garden was created by the Assyrian king Tiglat-Pileser I.

In the alluvial plain, a dynasty of Kassite origin had by then established their power over Babylonia. Its seat was the city of Babylon where a set of surgical instruments including scalpels and probes was found during excavations in the 12th century BCE Kassite levels. The clay tablets written in Babylon under

Kassite rule, and in Ashur during the last half of the second millennium, shed light on the process of standardising literary and scientific traditions, and on the creation of Akkadian cuneiform books.

The 1ˢᵗ millennium BCE

The regions over which Assyria and Babylonia exercised their political power and influence were separated by a natural border: Assyria ruled over the area from the upper to the middle courses of the Tigris and Euphrates rivers, and Babylonia from their middle courses to the Persian Gulf.

This picture changed in the first half of the 1ˢᵗ millennium BCE. Assyrian kings in the north developed a dominant political position over all Mesopotamia during three centuries from 912 to 612 BCE. Under Esarhaddon (681–669 BCE), the Assyrian empire reached its largest territorial extension, including all of Mesopotamia, to the west Cilicia on Turkey's south coast, to the north Urartu in the Caucasian Mountains, to the east and southeast Elam, and to the south Egypt as far as Nubia. His son Ashurbanipal (668–631 BCE) inherited this huge empire. Both kings are intimately associated with the foundation of an extensive cuneiform library at Nineveh, usually referred to as Ashurbanipal's library. Most of the tablets found there are now held by the British Museum. Information about ancient Mesopotamian medicinal plants is found in different forms of cuneiform books kept there, such as herbals, copies of the *Uruanna* plant glossary and medical prescriptions. Ashurbanipal's library housed the largest collection of scholarly cuneiform clay tablets, but it was not the only library in Assyria or Babylonia, and lacked important resources such as the plant description texts known from Huzirina and other sites. Other important libraries have been excavated in Huzirina (Sultantepe), Ashur (Qal'at Sharqat), Kalhu (Nimrud), Babylon and Sippar (Tell-ed-Der). Ashur is also the location of a unique cuneiform tablet that records the medicinal ingredients that were stored in the private pharmacy belonging to a family of healing practitioners.

The Assyrian kings struggled to hold their empire together, and repeatedly had to put down rebellions in Babylonia or to fight against Egyptian, Elamite or Urartian armies. For some periods, Aramaic-speaking Chaldean dynasties ruled over parts of southern Babylonia. One of their kings was Marduk-apla-iddina II (722–710 BCE), the Biblical Merodach-Baladan. According to a small cuneiform tablet listing plant names, the king had a garden tract adjacent to his palace in which medicinal plants were cultivated. It was surely a modest plot compared to the enormous parks of the Assyrian kings but represents the first written account in history of a physic garden. The Assyrian empire eventually collapsed under the assault of Babylonian and Median armies. First Nimrud was destroyed in 614 BCE, and two years later in 612 BCE, Ashur and Nineveh fell. From 612 to 539 BCE, a Chaldean dynasty reigned in Mesopotamia, filling the power vacuum left by the Assyrian kings. The longest-ruling king of the dynasty was Nebuchadnezzar II (604–562 BCE). The Assyrian and Babylonian empires anticipate the consolidation of imperial hegemonies in the ancient Near East. Persians and Greeks with their mighty armies conquered the extensive territories formerly controlled by Assyrian and Babylonian kings and further expanded the bounds of the ancient world, which had strung out between the valleys of Mesopotamia and the Nile, from western Anatolia to the Indus river, and from Afghanistan to southern Egypt and Libya in North Africa.

The Persian and Greek empires

The second half of the 1st millennium BCE marks the end of ruling native-Mesopotamian dynasties and the end of Mesopotamia's political independence. In 539 BCE, Cyrus the Great (600–530 BCE), the founder of the Achaemenid Empire, invaded Babylon, heralding Persian power. During the rule of Xerxes I (486–465 BCE) and especially his son Artaxerxes I Longimanus (465–424 BCE), the Greek historian and geographer Herodotus (c. 485–c. 425 BCE) travelled across the Mediterranean world. The accounts of his travels are included in the nine books of his *Histories*. Although he spent some time in places such as Egypt or Samos, becoming familiar with local customs, it is not clear whether Herodotus actually visited Babylon to see at first hand the Tower of Babel and the impressive city walls or whether he relied on distorted oral tradition.

The Achaemenids ruled over Mesopotamia until Alexander the Great's Macedonian army entered Babylon in 330 BCE. The Hellenistic or Seleucid period (330–141 BCE) is, as the name indicates, associated with the Greek control of Mesopotamia. During Alexander's reign and that of his successor as satrap of Babylon, Seleucus I, the Greek philosopher Theophrastus (c. 371–287 BCE) wrote his *Enquiries into Plants*, which includes many references to the vegetation of Mesopotamia, such as a detailed description of the cultivation of date palms in Babylonia sent by one of his students.

Mesopotamia passed into Parthian or Arsacid hands, a Persian dynasty that ruled from 141 BCE to 224 CE. During the reign of Mithridates II (124–91 BCE), the

Arsacid dynasty extended its power over Mesopotamia and Armenia, marking the start of contacts with the Roman Republic. The Greek physician Dioscorides (c. 40–90 CE) wrote a highly influential five-volume book *De materia medica*, drawing on his experience as a physician with the Roman army. Many of the uses for medicinal plants that he describes can be found already in the cuneiform plant lore.

In Mesopotamia, the cultural life of Babylonians did not cease under foreign rule; the care and preservation of the millennial cuneiform tradition was maintained but restricted to the temples in the cities of Babylon and Uruk, where the study of astronomy and astrology continued. It is thus not a surprise that the latest datable cuneiform tablet written in 80 CE is an astronomical almanac. A small group of clay tablets from Babylon dating to the 1st century CE and using the Greek alphabet to write Sumerian and Akkadian words marks the end of the cuneiform script, but Babylonian knowledge did not die out. As discussed in Chapter 5, ancient Mesopotamian medical practices may have filtered through to authors writing in the Greek, Syriac and Arabic languages over the following millennium.

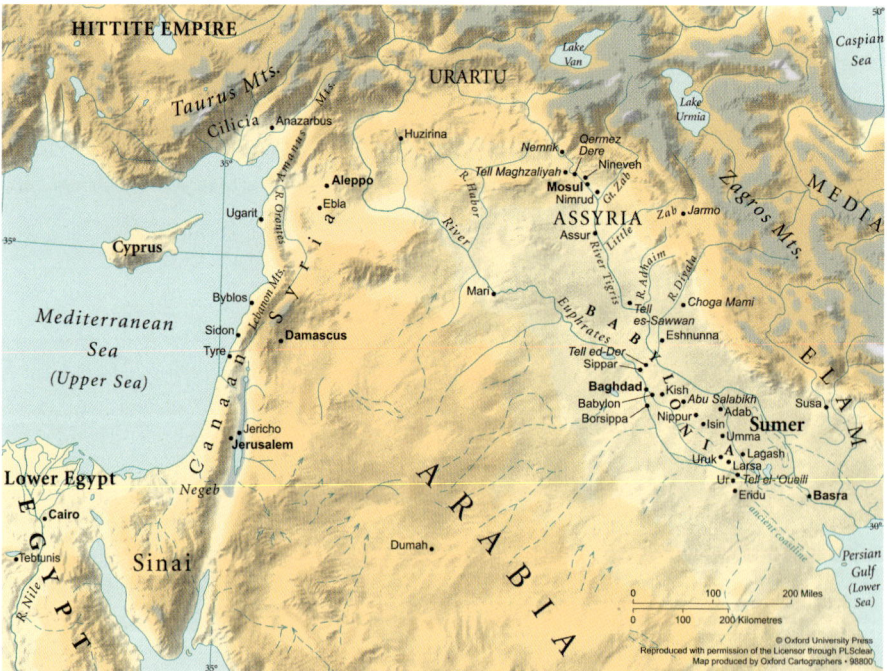

Map of Mesopotamia, the region within the Tigris and Euphrates river system
© Oxford University Press, reproduced with the permission of the Licensor through PLSclear, map produced by Oxford Cartographers 98800

2: SEEKING HERBS AND ROOTS

Nothing gives a more vivid insight into the medical care at a royal court in 1st-millennium BCE Mesopotamia than an episode in the final years of the Assyrian king Esarhaddon (born c. 715– died 669 BCE). The two principal protagonists in this chronicle are the chief physician Urad-Nanaya, trained in medicinal plant lore, and the chief exorcist Marduk-shakin-shumi, who specialised in the performance of healing rituals.

King Esarhaddon did not enjoy good health in general but when he caught a bad cold in the summer months of 672 BCE he became rather worried. Despite the reassurances of his chief exorcist that this was a seasonal flu and that the king would recover soon, the king suspected supernatural causes behind his worsening state of health. From at least November 672 to February 671, he ordered the performance of a number of healing rituals. One year later, he fell sick again and told his chief exorcist Marduk-shakin-shumi "my legs and arms are without strength. I cannot open my eyes and I feel very weak." On the basis of Marduk-shakin-shumi's diagnosis, the chief physician Urad-Nanaya sent a fever lotion to the king, telling him "surely the fever will leave. I have prepared this lotion already two or three times for the king – so the king knows it. If he wishes he should apply it repeatedly from tomorrow on and then his illness will recede." It seemed though that Marduk-shakin-shumi was doubtful about the true cause of this recurrent fever and told the king "this illness does not make much sense. Why should the king have caught it, it is early summer," and concluded that "this is the work of the gods". Urad-Nanaya shared this concern and sent a special herb to the king "one that looks like the base of an earring, is really efficacious and very rare. It is good against witchcraft." The combined efforts of the exorcist and the physician finally led to the king's recovery in June 670. But one year later around the same time the king fell sick again, this time vomiting bile. The new attack deeply alarmed the chief physician Urad-Nanaya. Yet he sought to encourage the king, and told him in June 669 "as stated in the medical prescriptions manual 'if a man purges through mouth and anus, then he shall recover'". But nothing could be done; the worst fears of the two healing experts were confirmed when the king died months later on 1 November 669, aged 44.

It is hardly possible to identify the plant used by Urad-Nanaya and to find out how he acquired it but there is ample information about other plants that were farmed, grown in gardens or gathered. As for Esarhaddon's chief physician, practitioners like him or Marduk-shakin-shumi, serving at the royal court, were in a privileged position and had more extensive means of procuring medicines than the healers working in towns or in the countryside.

Practitioners and patients

The incident involving Urad-Nanaya and Marduk-shakin-shumi shows that there were two healing professions in ancient Mesopotamia; one is called in Akkadian *asû*, which is usually rendered "physician, pharmacist" and the other is *āšipu* (also called *mašmaššu*), the "exorcist" or "magician". Both Akkadian words belong to the large group of culture-specific terms and concepts that have no equivalent in English (or other modern languages) and thus cannot be exactly translated. The main difference between them is that the magician was authorised to expel disease demons with incantations and ceremonies of exorcism. Whether he acquired this power because of the study of supernatural powers or because of lineage and consecration is not clear. By contrast, the pharmacist recited only medico-magical spells and was not able to address evil demons.

The treatment of both healers consisted of a mixture of therapies and healing rituals. The spells were often recited during the preparation of the medicine to enhance its effect or during the treatment to calm and reassure the patient. As stated in many colophons, that is, the inscriptions at the end of a cuneiform tablet that include information about its production, physicians and exorcists copied, edited and consulted the same works about medicinal plants and medical recipes. Writing herbals and the *Uruanna* plant glossary formed part of the training of the two kinds of practitioner. A major part of the learning was probably based on oral instruction and systematic observation. One should not forget that medical education, like that in other professions, was embedded in the network of larger families whose members would usually pursue the same or similar lines of work; thus the apprentices would learn their craft within their families.

The *asû* and *āšipu* or *mašmaššu* had to be well versed in cuneiform writing and reading to be able to consult, for example, the manual of medical recipes, which in the city of Ashur was kept in the temple library. Exorcists of the city would excerpt specific recipes in the library and take them home to prepare the medication. The written tradition, which contains predominantly information about the elite class, usually refers to doctors who practised at the palace or provincial courts. They could be sent from there to make home visits and to attend sick dignitaries and their families. It seems that healers without a fixed workplace would travel across the country to offer their skills to the sick for payment. We do not know how high the fees were but we know that some healers wanted to make more money because incantations have been preserved that were recited to increase the income of 'exorcists'.

The practitioner is described here as 'he' because there is hardly any evidence that women practised medicine. However, women midwives existed and, then as now, surely village women knew where to gather healing plants and had an important role in healing. Their work has left few traces in the written record.

Physic gardens

At least one king is known to have had a physic garden (Akkadian *gannatu*), namely Marduk-apla-iddina II (721–710 BCE), the Biblical Merodach-Baladan. He belonged to the large Chaldean tribe of Bit-Yakin; his power reached from the Arabian Peninsula to southern Mesopotamia, where he became a fierce adversary of the Assyrian kings. Information about this garden comes from a cuneiform clay tablet kept at the British Museum that lists more than 60 plant names. The names are arranged in 15 sections with groups of between three to seven plants. It cannot be determined whether this division refers to the actual garden design, for example the space the plants need to grow, or whether it represents a classification system. Among the listed plants are garlic, onion and

Cuneiform clay tablet with a list of the plants that grew in Marduk-apla-iddina's physic garden, c. 7ᵗʰ century BC
© The British Museum, BM 46226

leek, possibly turnip, radish, gourd and melon, and leafy vegetables and spices such as coriander. All are typical food plants in the ancient Mesopotamian diet, so the cultivated plot is usually interpreted as a kitchen garden used by the royal cooks. But food plants were also used as drugs. A closer look at the context in which the plants appear reveals that one third are described only in medical treatises, one third are mentioned in both food-related and pharmacological writings, and for one third there is not sufficient information on whether the plants were consumed as food or prescribed as a drug. It cannot, of course, be ruled out that the kitchen staff of Marduk-apla-iddina's chef went to the garden to fetch some vegetables and herbs, but it is more likely that the palace kitchen had its dedicated food supplies. Though the list does not include poisonous plants, it does mention plants with a strong purgative effect and rather bitter taste. To distinguish between everyday food and medicinal ingredients was the responsibility of the physician trained in herbal lore. This small clay tablet is thus the first evidence in history of a physic garden. A court physician with access to such a garden could probably have obtained there some of the plants discussed in this book, including onion, garlic, leek, garden rocket, greater plantain or ribwort, and coriander.

Gathering from the grave

We do not know whether the plant prescribed by Urad-Nanaya was so rare because it was uncommon in the wild, or was imported from far away, or whether he had gathered it in some special place. One of these special places where practitioners occasionally gathered plants was the grave. False carob and camel thorn had to be picked at times at the resting place of the dead, and it is not surprising that they were among the typical ingredients for alleviating those pains for which the spirit-of-the-dead was held responsible. False carob has very long taproots that can grow more than 15m deep. Mesopotamians believed that they would reach the underworld where the ghosts of the dead dwelt and therefore be in contact with them. The best time to fetch not only these two plants, but also greater plantain or ribwort, was at night. The practitioner had to leave a gratuity under the false carob and follow precise directions: *"If a woman suffers from complications during delivery, gather shoots, heap them under the false carob that grows near a mudbrick wall and say 'accept your present and give me the drug of life, so that the woman gives quickly birth!' As soon as you have said this, pull out the root and crown of false carob, go and do not look behind you and speak to nobody. Then twine a cord, tie one end of it around the plant parts, the other on her left thigh and she shall recover."* [1] Such instructions are unusual and specific to magical healing. The shoots are given to the plant in exchange for picking parts of it, in the expectation that the world of the dead would ensure a healthy birth. The cord tied around the leg and the plant possibly represent the umbilical cord and the infant. The false carob probably did not die from the chopping of crown and root because the plant would regrow from any roots the practitioner left in the ground.

Plant stores

Shops where a physician could buy medicines did not exist in ancient Mesopotamia. The earliest recorded pharmacies opened during the 8th century CE in Baghdad under the Abbasid Caliphate. Some healers are known to have built up their own stock of medicines. Little is known about such storehouses in ancient Mesopotamia; some seem to have been attached to temples, others to palaces. One of the best-preserved documents about a stock of medicines comes from 8th-century BCE Ashur. In this northern Mesopotamian city lived a prominent and wealthy family of practitioners – physicians and exorcists – who worked for private individuals and official institutions. Among the cuneiform clay tablets that made up their extensive library is a document that lists about

Cuneiform clay tablet with a list of medicinal ingredients kept in the private storehouse of a family of healers, Ashur, c. 8th century BCE
© Staatliche Museen zu Berlin, Vorderasiatisches Museum, VAT 8903, Photo: Olaf M. Teßmer

230 medicinal ingredients; around 190 items are plants or plant products and 40 ingredients are of mineral and animal origin. Since the document is unique it seems likely that only a few practitioners were able to build up such stores. The tablet was written by an assistant whose task was to check the contents of the store, recording how the ingredients were kept and where the plants came from. The plants are said to come "from the open country", a designation for wild plants. This suggests that the practitioners themselves went in search of wild plants to meadows at forest margins, mountains slopes and riverbanks, or that they had the plants picked by their assistants trained in herbal lore.

A handful of known orders issued by practitioners, and receipts for medicinal plants made out by them, suggest that some could order their remedies directly from gardeners or from temple and palace storehouses. This service was only for those professionals who treated the elite of society in the royal and provincial courts. One of these receipts contains the ingredients for a salve a practitioner prepared for the children of the Assyrian king Assur-bel-kala (1074–1056 BCE). Among the plants are red bryony, juniper and tamarisk seed. Widely cultivated plants such as cumin, coriander, garlic, onion, leek and dates must have been within easy reach of the physician.

We do not know to what extent medicinal plants were traded. Sweet flag, for example, does not grow in Mesopotamia and must have been imported either from Central Asia in caravans or via sea trade from Arabia and India. Since at least the middle of the third millennium BCE, Mesopotamian rulers maintained trade relations with the Indus Valley, Oman, Bahrain, Iran, Pakistan and Baluchistan. Among the typical trading goods were precious stones and metals, sissoo or Indian rosewood (*Dalbergia sissoo*), which was highly valued for construction, and exotic animals.

Preservation

Plants were preserved by hanging on wooden herb-drying racks. As stated on the tablet from 8th-century Ashur described above, the store held a rack with four bars; each bar held about 20 plant bunches. Those plants that could not be fixed to a rack were kept in typical household dishes such as pans (Akkadian *qālītu*) and jars (Akkadian *qabūtu*). As for the plants discussed in this book, greater plantain or ribwort, liquorice and nightshade were dried on the first bar, and on the second hung henbane and tamarisk branches. Although unspecified, the following plants were in all likelihood stored in some sort of container, possibly after drying: bryony, juniper, colocynth, camel thorn shoots,

false carob shoots and linseed. Cedar, sweet flag, juniper berries, juniper berry pulp and dates were kept in clay pans that could be used for roasting. Although there is no information about the size or shape of the pan or whether they were open or closed, a small drawing incised on the edge of a clay tablet gives an image of a *qabūtu* jar. This kind of container was used to store cedar tar.

One of the few drawings on cuneiform clay tablets; it shows a *qabūtu* jar incised on the right edge of the tablet. Babylon, 550 BCE
© The British Museum, BM 63869

Reconstruction of racks for drying herbs
© Barbara Böck

3: ADMINISTERING HERBAL MEDICINE

Plant products such as leaves, roots, seeds, pulps, milky saps, resins, barks or peels were the most used ingredients. Recipes occasionally record whether the plant or some of its parts had to be fresh or dried. Mineral substances and ingredients of animal origin were also employed but far less often. Curiously enough, hot infusions or herbal teas, the typical format of household remedies nowadays, seem not to have been used by Babylonian and Assyrian practitioners.

Medicinal drinks and foodstuffs

A herbal potion was the most common way to take medicine. Usually the medicinal plant was first crushed, then macerated or boiled. Only a few recipes prescribe the use of one ingredient; more often, a medicine contained more than one component. The basic carrier liquid of choice for the medicine was beer of different brewing qualities. The consumption of beer, made since prehistoric times in the valley of the Tigris and Euphrates, was a daily necessity because of the scarcity of safe drinking water. Owing to its alcohol content and low pH value, beer is a poor habitat for microorganisms. If water had to be used it was previously boiled and, if required, mixed with herbs. Other liquid carriers were oil, syrup, wine, vinegar and milk.

The ancient Mesopotamian practitioners distinguished between two basic extraction methods, namely maceration and decoction. Maceration requires that drugs are in contact with the liquid for a longer period of time, usually at room temperature (15°–20°C). Babylonians and Assyrians did not employ the expression "room temperature" but they had a clear understanding of the ambient temperature and times required. What they did was to place the medicine overnight on the roofs of their houses. Sometimes they would add to the recipe the remark that the mixture was to be exposed to the starry sky, preferably, when the "Goat star" had risen. The "Goat star" corresponds nowadays to Lyra, a small constellation situated north of the celestial equator, which ancient Mesopotamians saw as a manifestation of their Healing Goddess Gula. This exposure or irradiation by her stars magically guaranteed and enhanced the efficacy of the medicine, serving at the same time to macerate the drugs. Another form of maceration was to warm up the mixture at low temperature. In modern terms, a temperature of 40°–50°C is recommended. Ancient Mesopotamians enclosed the medicine in tannours or cob ovens, both made of clay, profiting from the heat that was left inside after baking and cooking food. This way of preparing the medicine was quicker and shortened the whole process.

For the second main extraction method, decoction, the ingredients were added to a liquid and brought to a boil. After the mixture had cooled off, it was filtered and the potion could then be drunk.

Another, simpler and quick way was eating the medicinal plant without previous preparation. In these cases the whole plant or its leaves and roots were consumed fresh. Thus, for example, the recommendation to swallow 14 garlic cloves for recurrent colic and intestinal cramps.

On the importance of beer

Beer was an indispensable foodstuff in Mesopotamia. It was brewed with barley malt, to which occasionally emmer wheat (*Triticum dicoccum*) and barley coarse meal were added. Spices and herbs could be mixed in to enhance the bitter flavour. Beer was not only the principal beverage of the common man but was also served at the banquets of kings and used for libations to honour the gods.

Both barley and emmer wheat are cereals, in other words, members of the grass family (Poaceae) cultivated for their edible, starch-rich grains. Wheat and barley resemble each other in bearing their grains in an ear (spike) at the end of stalks, but there are differences in detail. Barley and emmer wheat are among the earliest crops taken into cultivation in the Near East. Wild barley and emmer both grow on the rocky hillsides that flank Mesopotamia, in an arc stretching from the Levant (Israel, Palestine and Jordan), through Syria, Turkey and Iraq, to Iran. Grains of wild wheat and barley have been found at archaeological sites in this region dating back to over 20,000 years BCE. Cultivation of wild grains may have begun around 10,000 years BCE, perhaps in response to climate changes following the end of the last Ice Age. Fully domesticated wheat, barley, lentils and peas occur from 8500 years BCE. Elaborate systems of irrigation canals and the first large cities developed side-by-side in lowland Mesopotamia during the period 4000–3000 years BCE. Archaeobotanical finds of cereal grain from ancient Mesopotamia, at sites such as Abu Salabikh (3200–2300 BCE), are dominated by hulled barley, with much smaller amounts of emmer wheat and free-threshing wheat (durum and bread wheat) present. Barley is an undemanding cereal and thrives well on the slightly saline soils that are so typical in Mesopotamia – a result of the salt content of irrigation water and the shallow saline groundwater.

Beer, as a liquid product, is naturally difficult enough to detect in the archaeological record. Images on cylinder seals dating to the third millennium BCE show people sitting around a large vessel, drinking unfiltered beer through long straws.

Chemical residues that may derive from beer are found on pottery vessels from the fourth millennium onwards. However, neither these, nor the remains of breweries found in several excavations, are sufficient to clarify a brewing process that is hard to disentangle in the written record. It is likely that malted barley was combined with a ground barley product (sometimes incorrectly identified by scholars as bread), possibly in a process analogous to that proposed for ancient Egypt where analysis of ancient brewing residues suggests that malted cereals were mixed with cooked, cracked grains. A similar process is used in some traditional African beers. Such a beer would be very different in flavour from a modern European beer.

Both Assyrian and Babylonian healers prescribed beer as a single ingredient. In case of an affliction of the hips, flavoured beer was recommended: "*In order to remove the disease called* saššaṭu *which affects the joints and tendons of the hips, have the patient drink repeatedly flavoured beer, then he shall recover.*" [1]

Beer is a highly nutritious source of vitamins (particularly B vitamins), minerals (particularly selenium and silicon) and dietary fibre. As a bitter drink, it also stimulates the appetite. It seems that practitioners recommend beer to malnourished people because of these health-giving effects: "*If the person is undernourished due to famine, place 1 litre of first-quality beer on the roof under the starry sky and have him drink it the next morning on an empty stomach."* [2]

The Greek physician Dioscorides (II, 87) mentions that beer is a diuretic and affects kidneys and tendons, but warns against its consumption because it produces excessive gas and is harmful to the kidneys and bladder. The Arab physician Ibn al-Bayṭār (no. 1689) observes that beer can strengthen the stomach provided it is flavoured with spices.

Pills and suppositories

Pills and suppositories were made of suet, beeswax, ghee, date pith, pulp of colocynth and several tree resins. Ancient Mesopotamian practitioners distinguished three different sizes: the small "pill" (expressed with the Akkadian verb *kapātu* "to roll into a pill"), the medium size called "acorn" (in Akkadian *allānu*), and the large size called "finger" (in Akkadian *ubānu*).

The basic substance for suppositories was suet, previously extracted from the adipose tissue of the kidneys and loins of slaughtered beef and mutton. The suet could be mixed with wax, plant resin or date pith. The ingredients for pills and suppositories had to be crushed and pounded very fine before being added. The pills were swallowed; the suppositories or pessaries were applied, respectively, into the anus and vagina.

Powders

Powders also required that the ingredients were very finely crushed. They could be taken pure or were mixed with flours of roasted grains, wheat, malt or copper dust. The fine particles of plants were used externally to treat wounds or skin irritations. When internal use was recommended, the powder had to be introduced with the help of a bronze tube into body cavities such as the ears, nostrils, eyes, vagina and penis. Occasionally, the patient had to snuff or inhale the powder through the nostrils.

The degree of fineness of the ingredients depended on the form of application. The treatment of gangrene of the feet, for example, included sprinkling the coarsely crushed leaves of colocynth on the aching parts.

Enemas and catheters

Enemas were usually administered with a solution based on beer, syrup, oil or water. Ancient Mesopotamian practitioners used two types of instruments to apply them: one is a tool in the shape of a flute made out of reed and wrapped in cloth; the other is a leather hose. Some of the prescriptions for enemas recommend the processing of an unusually high number of drugs amounting to nearly 100. Their intended effect depended on whether the enema was meant to treat constipation, relieve inflammations, destroy parasites or provide relief from gas. Enemas and potions were the two most important methods of treating the sick.

Some urinary diseases were treated by inserting liquid into the bladder with a small copper tube. These ailments included, among others, bladder stones and dribbling urine, which were treated with an extraction of black nightshade.

Lotions, ointments, pastes and poultices

Lotions were prepared with a base of water, oil or syrup. They were meant for external application without using friction. After treatment, the affected area could be covered with a bandage. For example, in case of fever that spread to the eyes, juniper oil was dribbled onto the forehead and temples.

Ointments were used to soften or soothe the skin, whereas liniments were rubbed into the skin to treat muscle diseases and complaints, and were mostly based on oily carriers, including ghee.

Pastes differ from ointments and lotions in that they contain a higher proportion of finely powdered ingredients, which make the medicine stiff or even solid. The substances for pastes were mixed with oil, resins or plant pulps and then applied to the body with the help of wool or textile strips.

As for poultices, the medicinal ingredients were added to a hot or warm carrier such as beer, water, oil, syrup, vinegar, milk or wax. The mass was then spread on a dressing or applied directly to the affected areas such as eyes, head or limbs.

Most medicinal plants administered in the form of lotions etc. were understandably used to treat skin ailments, sores and wounds such as plantain for dog bites, but a small number were applied externally for coughs and digestive problems.

Dressings

Tufts of sheep wool were used to prepare absorbent dressings or pads; the fibres of the fleece were wrapped round the medicine. Recipes explain how these were employed to make fertility and pregnancy tests. Pads could also be used in the ear. If pus or blood was flowing from the ear, for example, healers would prepare a pad with camel thorn and insert it into the patient's ear canal.

Fumigations

Fumigation was another common form of applying medicines. The medicinal ingredients were put into a censer or burnt over charcoals. The patient then had either to inhale the fumes or expose parts of the naked body to the smoke. Fumigations were often prescribed for diseases that were believed to originate from supernatural agents such as demons, deities and spirits. The active substances in the drug enter the bloodstream more rapidly when inhaled and quickly trigger the desired effect.

Inhalation had other uses: priests and exorcists would breathe in juniper and cedar smoke before engaging in their ritual practices.

Medical instruments

Cuneiform texts hardly refer to medical instruments and surgical tools, occasionally mentioning copper tubes, small reeds, fine knives and scalpels. To alleviate the pain caused by eye afflictions, practitioners used bronze lancets to incise the temples. Blood-letting for treating eye diseases must have been

Set of surgical instruments found at Babylon containing bone spatula, as well as a probe, ear hook, mirror probe, ear probe, sharp spoon, and long hook. Babylon, c. 13th century BCE

© Staatliche Museen zu Berlin, Vorderasiatisches Museum, VA Bab 4108.001, VA Bab 7584, VA Bab 7587, VA Bab 7570, VA Bab 7562, VA Bab 7562, VA Bab 7594, Photo: Olaf M. Teßmer

common because it is mentioned in the laws of Hammurabi: "*If a physician (asû) performs surgery with a bronze lancet and heals the patient or opens the patient's temple and heals the eye, he shall take ten shekels of silver (as fee)*" (law 215). A number of instruments that have been found in the ancient city of Babylon show that physicians would examine the external ear cavity with scopes or probes and use needles, hooks and spatulas.

Measures and doses

Most recipes do not specify the amount needed, giving the impression that it was left to the experience and knowledge of the practitioner. Both weight and liquid measures were used from time to time. Indication of quantities less than 1g suggest the existence of beam balance scales.

Weight was measured in shekels, Akkadian *šiqlu* or Sumerian *gin$_2$*; one shekel corresponds to 8.3g. The liquid measure was called *qû* in Akkadian or *sila$_3$* in Sumerian; one *qû* is about 1 litre. The *qû* measure was further divided into one half and one third. One tenth of a *qû* was called *akalu*. When we quote from ancient prescriptions, quantities are translated into the metric system.

In a recipe for treating diseases of bile ¼ shekel (2.1g) of black bryony, a poisonous plant, was to be taken with 10 shekels (83g) water. For red bryony the following quantities are mentioned: 14 grains (0.7g) to treat intestinal worms, 1/6 shekel (1.4g) for jaundice and 15 grains (0.75g) taken in ½ *qû* (500ml) of sesame oil and beer for biliary diseases. As for poplar resin, a recipe for treating bloodshot eyes mentions the amount of 5 grains (0.2g). To help in case of a feverish head and bloodshot eyes, a poultice with 1/3 *qû* of fig leaves (0.3kg) was prepared. For sick eyes, 10 *kisal* (10.4g) of sesame oil were applied to the temples. And 10 shekels (83g) of dried sesame cake were required to prepare a poultice for fever and throbbing eyes.

Weight

1 shekel	8.3g
1 grain	0.05g (1/180 shekel)
1 *kisal*	1.04 (c. 22½ grain)

Volume

| 1 *qû* | 1 litre; 1kg |
| 1 *akalu* | 100ml; 100g (1/10 *qû*) |

4: CUNEIFORM SOURCES

Information about plants used for therapeutic purposes comes from four cuneiform genres: medical prescriptions, herbals (books of simples), plant description texts and plant term glossaries. Together with culinary texts and administrative records, these form the basis for matching Akkadian and Sumerian plant names to their present-day botanical equivalents.

Medical prescriptions

Medical prescriptions contain a wealth of information; with the exception of a few Sumerian medical recipes, they are all written in the Akkadian language. A typical prescription is structured as a conditional sentence. The if clause contains the description of the ailment from which the patient suffers, the name of the illness or the sick body organ; the result clause includes the instruction for the preparation and administration of the medicine and usually closes with the statement that the patient will recover. Each recipe includes a list of ingredients including plants, minerals and animal products. Few prescriptions give only one single substance (also called a simple); the great majority usually recommend a combination of between two and six products, and occasionally we find more ingredients. The patient is usually a 'he' (Akkadian *amēlu* "man"); only prescriptions dealing with problems occurring during pregnancy refer explicitly to a woman patient (Akkadian *sinništu* "woman").

Medical prescriptions were systematically collected and edited in books. Among the important find-places for medical prescriptions are the temple, palace and private libraries excavated in the Assyrian cities of Ashur, Kalhu and Huzirina and the Babylonian cities of Sippar, Borsippa, Babylon, Kish and Uruk. The Assyrian finds usually date between the 9th and 7th centuries BCE and the Babylonian a little later, between the 6th and 4th centuries; a few collections of prescriptions come from the beginning and middle of the 2nd millennium BCE. It seems that various books of medical prescriptions circulated in the north and south of Mesopotamia, the main difference being the internal structure and length of the chapters or tablets. For example, the version from the southern city of Uruk consisted of at least 45 chapters or tablets, whereas the edition from Assurbanipal's library in northern Nineveh comprised twice as many chapters. This Nineveh book is also the best-preserved and prescriptions following this edition can be found in other Assyrian libraries, such as the one in Nimrud (ancient Kalhu). All versions have in common the typical *a capite ad calcem* structure, that is, arranging the diseases according to the affected body part from head to toe. As with most cuneiform books, medical prescriptions cannot be attributed to authors. In contrast to later classical works, such as the ones authored by Theophrastus, Galen, Pliny or Dioscorides, we have practically no knowledge about who conceived these texts, and only know the names of the scribes who copied and recopied them. Many of these were practitioners but they did not sign as author in the sense of creator; rather, they understood themselves as editors who revised, reshaped and preserved material that has been handed down for centuries.

Translations by researchers are based on the original cuneiform clay tablets that have been deciphered sign by sign. When, as is often the case, tablets are fragmentary, gaps have been filled with text from comparable cuneiform tablets. Once the full text is established, the task of translation begins. The question is whether to choose a literal translation, maintaining as far as possible the structure and literal meaning of the Akkadian words, or a functional, meaning-based translation. Assyriologists usually render medical texts word-for-word, including when the ancient practitioners use a technical language

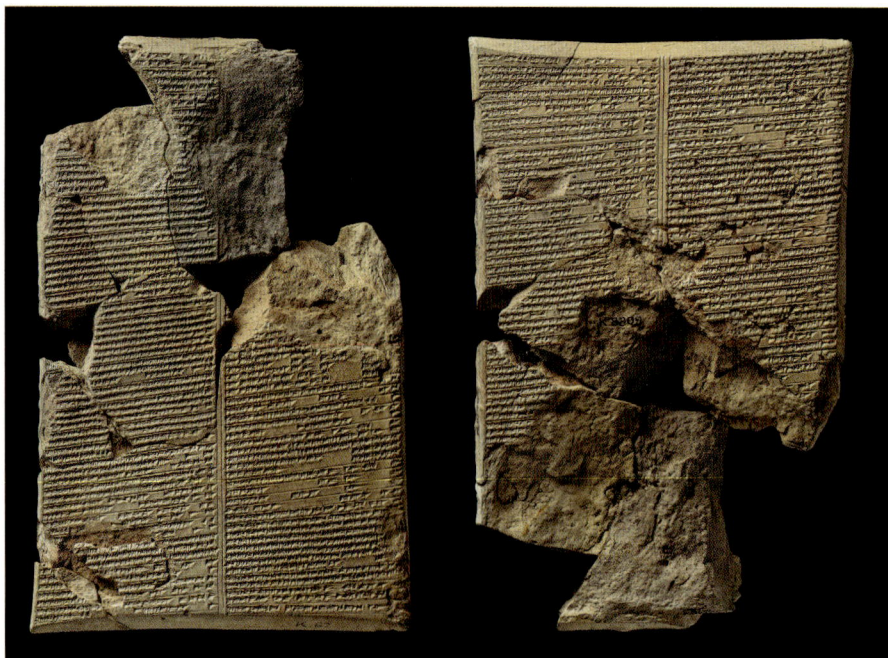

Cuneiform clay tablet with chapter 4 of the Nineveh edition of medical recipes on diseases affecting the hamstring muscles. Library of Ashurbanipal, 7th century BCE
© The British Museum, K.67

or idiomatic expressions. The result is often barely comprehensible in English. Take one example of such a translation. A formal and literal translation of the first line of the book of medical prescriptions is "If a man regarding the upper part of his skull contains heat," and one of the prescriptions to treat this heat is "you shall dry sesame residue, grind it, sieve it, knead a dough with mustard water and bind it". By contrast, the natural rendering is *"If a man's forehead feels hot, grind and sift dried sesame residue, make a poultice with mustard water and bind his head."* [3]

The translation of medical prescriptions is in progress, but we are far from forming a complete picture; many medical texts are kept, still unpublished, in museums worldwide or are published in cuneiform hand-copy only. Medical recipes are scarcely studied; only a few publications deal with aspects of the preparation of medicines, and almost nothing is known about the specific vocabulary of different plant parts and plant products. Also, the internal structure of recipes, in other words the number of prescriptions for the same ailment, or the number of ingredients contained in a medication, has never been analysed. As a rule, prescriptions usually stipulate more than one ingredient; the number of prescriptions with one single ingredient (called a simple) are few. Yet to define the specific medicinal use of a plant, medical recipes recommending simples are the most relevant.

Herbals (books of simples)

Herbals are books of simples, medicaments that are composed of only one plant. The oldest example of a herbal dates back to the 24th century BCE and was written in a Semitic language related to Akkadian, Eblaite, named after the ancient city of Ebla located about 55km southwest of Aleppo in Syria. This small clay tablet contains just three entries.

The first herbals written in the Akkadian language date back to the last quarter of the 2nd millennium BCE; the majority come from libraries dating to the 1st millennium BCE in the cities of Sultantepe (ancient Huzirina), Nimrud (ancient Kalhu), Ashur, Nineveh, Kish, Babylon and Sippar. The herbals are only available in cuneiform text editions. The texts are usually organised in three columns. The first column gives the name of the simple, the second the indication for which ailment the plant is employed, and the third instructions for how to prepare and administer the medicine. There are two different editions of the three-column herbals: the first is organised according to the simple cited in the first column and the second column gives a list of all the illnesses that could be treated with that plant; the second group is arranged according to the ailments in the second column, while the first column lists all those plants that are used to treat that illness. The first type presumably served to teach the medicinal values of plants. The second type helped users to look up, rather quickly, all the drugs that were used to cure the same disease. To give an example of this type: *"'Fox grapes' (black nightshade) – medicinal plant for scorpion sting – give to drink in first-class beer, anoint with oil."* [2]

At least two herbals were copied by apprentices, which suggests that writing herbals formed part of the training of practitioners. In addition to the common three-column herbals, there is an edition with two columns that is only known from king Assurbanipal's library in Nineveh. Like the king's special edition of the *Uruanna* plant glossary, this version does not seem to have circulated outside Nineveh. However, the Nineveh library stocked copies of the two three-column herbals as well. The abridged edition drops the references to the preparation and administration and is arranged according to diseases, giving all known simples in the first column. One wonders whether the king or his trusted practitioner used these special editions to check the knowledge of the physicians and exorcists who belonged to the palace staff.

Fragment of a herbal, c. 5th century, Babylon. MM 501
Courtesy of the Museu de Montserrat, Barcelona, Photo: Barbara Böck

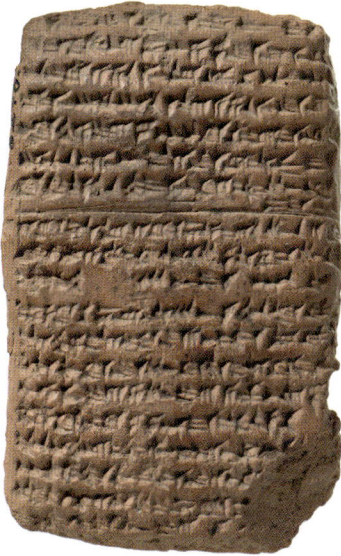

Cuneiform clay tablet with plant descriptions. Ashur, c. 8ᵗʰ century BCE
© Staatliche Museen zu Berlin, Vorderasiatisches Museum, VAT 8246, Photo: Olaf M. Teßmer

Plant description texts (*Šammu šikinšu*)

Šammu šikinšu, the opening words or title of plant description texts, means literally 'the appearance of the plant'. None of the 15 extant Akkadian manuscripts of the book are completely preserved; they all date to the first millennium BCE and come from libraries in Ashur, Nimrud (ancient Kalhu), Sultantepe (ancient Huzirina), Sippar and Babylon. The texts have been translated, but the plants and technical terms have not been explored. The treatise gives detailed descriptions of about 50 plants in comparative terms with other plants, just as we do today when describing a new species. The entry usually opens by comparing the plant's general appearance with that of another plant; then follows a description or comparison of the seed, leaf, fruit, root, flower and growth with that of other plants. Finally, the entry includes the ailments for which the plant was especially efficacious and a short indication as to how the medicine was to be prepared and administered. The value of such descriptions depends on how much is known about the plants that are used for comparison. Sometimes the description is so distinctive that it matches a specific plant that grows in modern Iraq. This is seen in the case of the plant *imḫur-ašra*, literally "it (i.e. the plant)-faced-twenty". The first part of the entry reads: *"The plant creeps over the ground like the colocynth, the leaves look like that of henbane, its berry is red. It is called 'it-faced-twenty' (black bryony, p. 63)."* [1]

45

Plant glossary (*Uruanna*)

Sometime in the second half of the 7th century BCE, the Assyrian king
Ashurbanipal decided to issue a new edition of the glossary of medicinal
ingredients, *Uruanna*. If we believe the statement the king left at the end of each
of the twelve chapters or tablets of the book, he wrote the glossary himself. He did
so, as he writes, because the older versions were outdated and incomplete. Since
the king had received a scholarly education and kept a private library in his palace
in Nineveh, it is possible that he was indeed intrigued by plant lore. Curiously,
however, he did not claim to have edited any other book in his library. Perhaps
his particular interest in medicinal plant names was due to his father's illness.
Assurbanipal was Esarhaddon's third or fourth eldest son and had seen in person
the deteriorating health of his father; he also must have seen how the therapies of
Esarhaddon's chief physician and chief exorcist failed again and again.

The title *Uruanna* comes from its first word, which is a plant name meaning
literally 'plant whose place is in heaven'. This plant is given as synonym for
the name *maštakal*, a word of unknown linguistic origin that is thought to be
soapwort – an identification that can neither be refuted nor confirmed. The
oldest version of the glossary has four chapters, dating from 12th to 11th centuries
BCE. Despite being recopied down to the middle of the 1st millennium BCE, the
Uruanna never achieved a standard edition. More than 50 incomplete copies
are known, almost all from Assyrian libraries in Ashur, Nineveh, Kalhu and
Huzirina. Like the herbals, the *Uruanna* glossary is available only in cuneiform
text editions. The glossary is structured in two columns with sections: horizontal
and vertical lines separate columns and sections thereby creating boxes. Each box
treats one plant and gives all the known names and spellings that were current in
different regions of Mesopotamia and occasionally names in foreign languages.
In its most complete version, the four-chapter edition had more than 1,700 line
entries. The purpose of such a work is evident – without agreeing on which
name refers to which plant, communication between practitioners risked being
hampered by ambiguity, eventually jeopardising the effect of plant remedies.

Some plant names refer to ailments, often providing information about the
specific properties of a plant lacking from herbals or medical prescriptions. The
names are formed with the general term for a plant, Akkadian *šammu*, to which
the illness, cause of pain or effect is attached such as 'plant for scorpion sting'
(Akkadian *šammi miḫiṣ zuqaqīpi*). That is the reason why the *Uruanna* glossary is
useful for identifying plants. To give an example, the plant called in Akkadian

Cuneiform clay tablet with chapter 3 of Ashurbanipal's edition of the *Uruanna* plant glossary. Library of Ashurbanipal, 7ᵗʰ century BCE
© The British Museum, K.4345

egengiru, identified as garden rocket, was not only known under the alternative Akkadian name *šurdunû* but also called 'plant for impotence' (Akkadian *šammi nīš libbi*) and 'plant for purging' (Akkadian *šammi šūšuri*). No prescription has been preserved that would recommend the use of garden rocket as single ingredient for treating problems of erection. Yet healers must have been familiar with the stimulating properties otherwise they would not have called it simply 'plant for impotence'. The same holds true for the reference to the laxative and emetic effect – it is not pointed out elsewhere.

Culinary texts

This book includes not only poisonous or inedible plants but also typical spices and other food plants. Information about their use and where they were cultivated comes from cooking recipes and administrative documents. Culinary recipes give insight into the Mesopotamian food landscape; we learn, for example, that coriander and cumin belonged to the typical spice mix that gave flavour to meat and vegetable dishes. Onion, leek and garlic were also standard ingredients, while colocynth seems to have been used to tenderise meat. The recipes are usually short, with much left to the imagination and knowledge of the actual cook. However, there are also more elaborate instructions. Here is an example of a short recipe: *"Broth with mustard – when fresh meat is not available use a dried one. Put it into water. Add lard, crushed mustard, onion, coriander, cumin, leek and garlic. Cover the pot and let the meat cook until it is tender."* [4]

Only a few culinary texts are known; three date from around the 18th century BCE, one about 1,000 years later. Occasional references to ceremonies such as a funeral suggest that some recipes were meant for special events and do not reflect everyday dishes.

Administrative records

Administrative documents in which the temple or the palace usually had a major role are the most informative source on agriculture and forestry. Details in these records can include a plant's habit and habitat, time of seed set and harvest, uses in construction, as raw material or its economic value. These records form the most commonly preserved cuneiform text genre, and the best studied. Transliterations and translations can be found in the Assyriological literature.

Key sources cited in this book

Source	Content	Note
Ancient Mesopotamia		
Administrative records	Uses and cultivation of a wide range of plants, particularly crops	Sumerian: 3rd mill. BCE; Akkadian: 3rd–2nd mill. BCE
Herbals (books of simples)	Plant names, ailments treated, and mode of preparation and administration. Arranged either by plant or by illness	Versions from Ebla (2400 BCE) to 1st mill. BCE; versions used here 8th–7th century
Medical prescriptions	Name of the illness or the sick body organ; preparation and administration of the medicine	Sumerian: end of 3rd mill. BCE Akkadian: 2nd–1st mill. BCE; Handbook of medical recipes from 7th century

Source	Content	Note
Literary texts	On the role of plants in the life and experience of Mesopotamians; symbolic connection between plants and man in rituals and incantations	Sumerian: 19th–17th century BCE; Akkadian: 19th–4th century BCE
Culinary texts	Recipes	Four known: 18th and 8th centuries BCE
Uruanna	Glossary of medicinal ingredients with all known names and spellings	Versions circulate from c. 1200 BCE; more than 50 copies known
Plant description texts	Detailed description of about 50 plants framed as comparisons	1st mill. BCE; 15 copies known

Greco-Roman world

Herodotus	*Histories* (written in Greek)	c. 484–425 BCE
Theophrastus	*Enquiries into Plants* (Greek)	c. 371–287 BCE
Dioscorides	*De materia medica* (Greek)	fl. 50–80 CE
Pliny the Elder	*Naturalis Historia* (Latin)	c. 23–79 CE
Galen of Pergamon	Many works, including *On the Properties of Foodstuffs* (Greek)	c. 130–200 CE

Medieval Arabic world

Abu Yūsuf Ya'qūb ibn 'Ishāq aṣ-Ṣabbāḥ al-Kindī (Al-Kindi)	*The Medical Formulary or Aqrābādhīn*	801–873 CE
Abū Ḥanīfah Aḥmad ibn Dāwūd Dīnawarī (Al- Dīnawarī)	*Book of Plants (Kitab al-Nabat)*	828–896 CE
Abu Rayhan al-Birūnī (Al-Birūnī)	*Book on the Pharmacopoeia of Medicine (Kitab al-saydala fī al-tibb)*	973–1048 CE
Abū-'Alī al-Ḥusayn ibn-'Abdallāh *Ibn-Sīnā* (Avicenna)	*The Canon of Medicine (Al-Qānūn fī'l-ṭibb)*	980–1037 CE
Diyā' al-Dīn Abū Muḥammad 'Abd Allāh ibn Aḥmad al-Mālaqī (Ibn al-Bayṭār)	*Compendium on Simple Medicaments and Foods (Kitāb al-Jāmi' li-Mufradāt al-Adwiya wa-l-Aghdhiya)*	1197–1248 CE

5: IDENTIFYING MESOPOTAMIAN HEALING PLANTS

The journey from an Akkadian or Sumerian plant name to its botanical identification is difficult, bridging a huge gap between ancient Mesopotamian and modern approaches to classifying nature. To achieve this, different bodies of knowledge are needed: of the ancient Mesopotamian tradition; the flora of Iraq and adjacent areas; modern and traditional uses of plants by herbal healers, and archaeological evidence for plant use. Thus, the research for this book required combined expertise in philology, botany, ethnobotany and archaeobotany. In this chapter, we explain more about the basis of each of these approaches.

Language

The identification of Akkadian plant names involves comparative linguistics, etymology and the study of texts. Linguistic studies compare the Akkadian word-root with related (cognate) languages. As Akkadian is a Semitic language, the same root is sought in related languages such as Aramaic, Hebrew, Syriac and Arabic. The translation of the same root in these languages is then applied to the Akkadian term. Since Sumerian is an isolated language without known cognates, no comparative linguistic studies for Sumerian plant terms can be carried out. Although this line of research provides important evidence, it does not take into account that the same root might not refer to the same plant. To give a modern example, the word corn, which derives from the Proto-Indo-European root *gre-no* "grain", refers to a single seed of a cereal plant or to plants that produce seeds. Corn has different meanings depending on the geographical region where it is used: English corn in England refers to wheat, in Scotland to oats, in the US to maize and German "Korn" to rye. All these cereals share the same root but mean different plants.

Common names evolve as part of patterns of thought. Indeed, as the study of the meaning of names or etymology shows, the common names of plants reflect their taste, smell, texture, touch, appearance and use. As a result, different species of plants can receive the same name simply because they share the same feature. However, when the linguistic origin of a plant name is unknown or not clear, no translation can be given. Not every culture categorises plants according to the same feature. To give a modern example, the English common names of the plant *Phedimus spurius* are 'dragon's blood' and 'stonecrop'. The first name refers to the leaves that change to brilliant red in autumn, the second to its habitat, a rocky location. The German common name *Teppich-Fettblatt* ('fat-leaf carpet') alludes to its mat-forming growth and texture of its leaves. The pattern of thought behind an ancient plant name only becomes evident once the plant is identified. Having identified the Akkadian plant name *lišān kalbi*, literally "Dog's tongue", as plantain we may assume that the plant may have received its name because of the similarity of its leaves to the tongue of dogs; or having identified the Sumerian plant name *še-du₁₀*, literally "sweet grain", as juniper we may assume that the tree received its name because of its fragrant berries.

Comparative linguistics and etymological studies form the starting point in the process of identification. Details about the nature of the plants come from the context in which the Akkadian (or Sumerian) name appears, viz. medical prescriptions, herbals, plant description texts, plant glossaries, culinary recipes or administrative documents. This information is then contrasted with the actual flora of Iraq, data about archaeological plant remains and the known medicinal uses of the possibly identified plants. To explain how the different bodies of knowledge are combined two examples are given.

The first is the Akkadian plant *šamaššammu*; the word-root can be compared to Arabic *simsim*, Hebrew *šumšum* and Aramaic *šumšā*, which all refer to sesame. The word travelled and entered the vocabulary of modern languages such as English. As for its etymology, the name consists of the elements *šaman* "oil" and *šammu* "plant", and literally means "oil-of-a-plant"; *šamanšammu* was then contracted to *šamaššammu* because it was easier to pronounce. The Sumerian name is *še-giš-i₃*, literally "oil-containing grain". According to etymology, the Akkadian and Sumerian names refer to an oil-producing plant, suggesting either flax or sesame, which were cultivated in ancient Mesopotamia. Sesame grows in Iraq, but because it is rarely found in the archaeological record, scholars discussed over many decades which plant the Sumerian and Akkadian terms referred to. According to cuneiform administrative documents that give the exact sowing and harvesting time *šamaššammu* or *še-giš-i₃* was a summer crop, ruling out identification with flax, which is grown as winter crops. Eventually, archaeobotanists were able to explain why ancient seeds of sesame are infrequently found: the oil-rich seeds become very fragile when charred, meaning that only few seeds have survived the course of time. The medicinal uses of the seeds of the plant named *šamaššammu*, as poultice to treat fever and inflammations of the skin and eyes, corroborate the identification with sesame.

The second example is the Akkadian plant *irrû*; no cognate terms can be found in Semitic languages. The Akkadian name means possibly "tangling (plant)", related to the Akkadian word for "tangle" *irru*. In Sumerian it is called *ukuš₂-ḫab*, literally "bitter *ukuš₂*". As Sumerian is not related to any other language, no comparative linguistic studies can be carried out. However, Sumerian *ukuš₂* is called in Akkadian *qiššû* and this term has cognates in other Semitic languages. They suggest a possible identification with gourd or melon. Akkadian *irrû* is one of the few plants for which a description is preserved; it is said that "its tendrils creep over the ground like those of the gourd or melon (Akkadian *qiššû*), its leaves are divided and its flower is yellow". According to medical prescriptions, the plant was prescribed as a strong laxative, administered to pregnant women and to treat skin ailments. Practitioners used the seeds, shoots, leaves, fruit, pulp and root. By triangulating all this information and comparing it with the actual flora of Iraq, we can propose colocynth as the identification. Further corroboration for this identification comes from comparing the ancient Mesopotamian medicinal uses with those in antiquity, medieval times and modern traditional medicine.

As can be seen, the identification of plant names is a process of matching and assessing the importance of diverse and complete data. Cuneiform tablets are affected by the vagaries of excavation, damage and publication. But even if all tablets were found and studied, they are highly selective in the information they record. As researchers we need to accept uncertainty in identifying plant names; in practice, we have differentiated between three levels of certainty: near certain (● ● ●), highly probable (● ● ○) and possible (● ○ ○). Sesame is an example for a certain identification, colocynth for a highly probable.

Comparing traditions

The ancient Mesopotamian herbals mark the beginning of the practice of listing *materia medica* with reference to their therapeutic use, anticipating the ancient Greek and medieval Islamic traditions. Ancient Mesopotamian practitioners, however, did not present their medicinal ingredients in alphabetical order. The *Uruanna* glossary is comparable to the lists of synonyms of simple drugs and of substitute drugs so common in Islamic pharmacology. In contrast to the cuneiform medical recipes, Greek and Islamic writers differentiated prescriptions according to their use of compound drugs or single ingredients. In spite of structural differences, we can find similar or even identical uses of plants in works by Graeco-Roman and Islamic scholars such as Pedanius

Dioscorides (b. c. 40 CE) or Abū Yūsuf Ya'qūb ibn 'Isḥāq aṣ-Ṣabbāḥ al-Kindī (b. c. 801 CE). The many parallels between the famous herbal of Dioscorides, *De materia medica*, and the cuneiform texts suggest that Mesopotamian lore was not lost but filtered through to the Graeco-Roman and Islamic worlds. Many of the uses in western herbalism today can be traced back to the writings of Dioscorides, and thus derive from the authority of the ancient Mesopotamian practitioners. We have therefore drawn on these later works (pages 48–9) for both plant names and traditional uses.

Although there are parallels between the cuneiform tradition and Greek as well as Arabic treatises, it is difficult to establish precisely whether or how Mesopotamian lore was handed down. The principal reason is that there are no written records available that would provide a similar basis for tracking a sequence in time of the transmission of knowledge, as there are for the circulation of Greek herbals in Syriac schools from the 6th to the 9th century CE and in Arabic schools from the 9th to the 13th century CE, or for Latin herbals in medieval Europe. Medicine was not learned in schools or academies in ancient Mesopotamia; rather medical education was embedded in the social network of sometimes extended families. Although some Assyrian and Babylonian cities were famous for their scholars, and for being home to great libraries, no institution existed that would have equalled the Alexandrian school in Egypt, which was one of the cultural hubs during the Hellenistic and Roman periods. As a remarkable centre of learning, Alexandria was the city where different cultural influences and tendencies in philosophy, medicine and literature mixed and were blended. Also, there is no evidence for the existence of schools of translation where the language and writing barrier from Akkadian to Aramaic (both languages coexisted during the 1st millennium BCE) or from clay tablet to papyrus or leather could have been crossed.

Even if there are no written traces, we know that cuneiform lore filtered through; this is evident from the many ancient Near Eastern mythological literary motifs that abound in the Bible, such as the story of the flood. Greek scholars and travellers visited Babylonian cities, a student of Theophrastus lived in Babylon, and Greek astronomers were well acquainted with the lunar theories of their Babylonian colleagues – all these examples reveal manifold cultural contacts. But unlike the written tradition, oral transmission is unstable and irregular, which makes it difficult to reconstruct lines of evidence.

Dioscorides' *De materia medica* and Galen's treatises *On the Powers of Simple Drugs, On the Composition of Drugs according to Types* and *Places* or *On* the Medicinal Properties of Foodstuffs were decisive for the development of the Syriac and Arabic pharmaceutical literature and accordingly for later medieval medical humanism. Syriac scholars such as the 6th-century CE Sergius of Resh'ayna produced a translation of Galen into Syriac, whereas the 9th-century CE physician Hunayn ibn Ishaq, who worked in Baghdad, translated the Greek into Arabic. Arabic pharmacologists considered Dioscorides and Galen as the greatest authorities. The Arabic translations were revised and distributed in the Eastern and Western Islamic world (in the west, via Cordova), through the work of important scholars such as Ibn-Djuldjul or Abu Rayhan al-Birūnī. Some of Hunayn ibn Ishaq's translations of Galen were translated from Arabic into Latin, as was Ibn-Sīnā's The Canon of Medicine; these and other translated works were studied in European medical schools such as the one in 10th-century Salerno, the southern Italian port renowned for its physicians. From Italy and Spain, the Greek and Islamic traditions spread further north to other European cities. Other herbals flourished and were widely distributed, as was the case, for example, the 12th-century Circa Instans, known as Book of Simple Medicines, which was translated into French, or the Old English Herbarium, an 11th-century translation of a 5th-century Latin treatise.

The many herbals produced in ancient and medieval cultures have in common that they were meant for learning, teaching and consulting. This explains why many herbals are illustrated – following the adage "a picture is worth a thousand words". The Roman naturalist and philosopher Pliny (Book XXV, 4) was the first to refer to medical authors who provided illustrations to their works. The oldest botanical drawings preserved are from the so-called Tebtunis herbal, a fragmented papyrus dating from the 1st to 2nd centuries BCE that comes from the ancient city of Tebtunis in Fayoum Egypt. Possibly, Dioscorides himself included original drawings in his herbal or had access to an illustrated herbal by Cratevus. Most famous, however, is the lavishly illustrated 6th-century CE Vienna Dioscorides, and some of these paintings are reproduced in this book (pages 70, 84, 139, 166, 172). These, together with other pictures, complement the few ancient Mesopotamian depictions of plants that have been preserved. As far as we know, Assyrian and Babylonian practitioners did not attach drawings to their herbals. When plants are represented in art they usually serve to characterise a landscape, as is the case in the reliefs that adorned the palaces of Assyrian kings. In these, the plants were carefully chosen either to identify the places and regions the kings had conquered or to boast of the opulent vegetation of their gardens.

Plants

At the heart of this book is a matching exercise between plant names in the cuneiform record, extracted from the sources described in the preceding chapter, and those plants actually available in ancient times, whether locally or through import. A good starting point is a list of plants growing in the region of Mesopotamia today, modified by evidence from ecology and archaeology as to which were growing there 4,000 years ago. But botany has far more to contribute. The ancient texts are full of information on the habitats of plants, their appearance and other properties; these can help to narrow down the range of species and, sometimes, to confirm identifications.

Although information on medicinal plants of the Middle East was given in the classical and medieval works of Dioscorides, Galen, Ibn Sīnā (Avicenna), al-Rāzī (Rhazes) and Ibn al-Bayṭār, it was not until the 16th century that the first research collection of medicinal plant specimens was made from Iraq. Leonhart Rauwolf, a German physician, botanist and a traveller, visited

(Left) Herbarium specimen of *Populus euphratica*, collected by Rauwolf by the Euphrates river, en route to Baghdad (Right) Illustration based on same specimen, in *Viertes Kreutterbuech – darein vil schoene und frembde Kreutter* (Rauwolf, 1583)

(Left) Courtesy: Naturalis Biodiversity Center, Leiden, specimen L.2111452. (Right) Courtesy: Biodiversity Heritage Library

Tripoli, Syria, Iraq and Palestine in 1574–5 to search for medicinal herbs that could be traded through his brother-in-law's firm. Shortly after his return, he published the results of his botanic expeditions in *Viertes Kreutterbuech – darein vil schoene und frembde Kreutter* [*Fourth Book on Herbs – Containing Many Beautiful and Exotic Herbs*] (1583). His herbarium is preserved at the Naturalis Biodiversity Centre, Leiden.

During the next three centuries, other pharmacists and botanists, including J.F. Gronovius, P.R.M. Aucher-Éloy, F.W. Noë and H.C. Haussknecht, travelled to Iraq and adjacent countries collecting plants. Their collections, with others, were enumerated in Boissier's *Flora Orientalis* (1867–88), still a useful reference to the plants of southwest Asia. After this, J.F. Bornmüller, F. Nábělek and H. Handel-Mazzetti made important contributions to the knowledge of the flora of Iraq, but it was not until 1929–33 that Evan Guest, in collaboration with the Royal Botanic Gardens, Kew, created the Rustam Herbarium, the first national herbarium in Iraq. After the Second World War, the Rustam collection moved to the Ministry of Agriculture's research station at Abu Ghraib near Baghdad. It grew steadily, with several Iraqi collectors collecting extensively in all regions of the country. Once the collections were believed to be exhaustive and to represent the country's flora, Ali Al-Rawi, the then Director of the National Herbarium, and Evan Guest started writing the comprehensive *Flora of Iraq*. Planned in nine volumes, they published vols 1 & 2 in 1966; then from 1968 to 1985 vols 3, 4, 8, 9 were published, edited by C.C. Townsend at the Royal Botanic Gardens, Kew. From 2013 onwards, Shahina A. Ghazanfar and John Edmondson took over the editing and completion of the remaining volumes. During this time, with collaboration and contributions from Iraqi botanists, three more volumes have been published. Work on the final volume is near completion, with Ali Haloob, a young Iraqi botanist and Curator of the National Herbarium, as one of the editors. Botanical research is well-established in Iraq, with several new herbaria established in the 21st century.

The *Flora of Iraq* is mainly based on the important collections made over the past century preserved at the National Herbarium in Baghdad and the herbarium at the Royal Botanic Gardens, Kew. Herbaria at Geneva, Paris and Vienna also hold collections from Iraq. These sheets of pressed plants are not only evidence of the plants present in the country, but also provide important information on their distribution, habitat, flowering time and vernacular names, making each sheet a valuable reference representing multiple data points.

Archaeobotany

There are several kinds of evidence that can be used to determine which plants were used in the past, and how, beyond the written sources discussed above. For example, pictures of plants appear on sculptured reliefs or engraved into seals; however, it is rare that we can identify anything beyond the most distinctive plants, such as date palms. In fact, the most informative evidence has proved to be the remains of plants, such as seeds and wood, found preserved in archaeological sites. When plant remains become charred, through contact with fire, they retain their shape but are resistant to the animals and microorganisms that usually ensure the decay of plant materials. Mesopotamia is not totally arid, so plant remains are not usually preserved by desiccation as they are in ancient Egypt.

The practice of archaeobotany, or palaeoethnobotany as it is sometimes called, is not straightforward. The first step is to collect the plant remains during an archaeological excavation. Sometimes fortunate circumstances (for us!) have led to the destruction of an ancient settlement by fire. Here one can usually find heaps of grain and other food plants, and materials such as matting, burnt where they were in use at the time of the fire. But such events are rare. More often, archaeobotanists work with ancient refuse. Charred plant remains from kitchen accidents, or crop by-products, wood and dung that were used as fuel,

Flotation machine in operation at an archaeological site. Excavated soil is poured into the water; seeds and charcoal float through the spout and into a sieve
Courtesy: Mark Nesbitt

were tossed into rubbish pits or middens. Since the late 1960s, the use of the flotation machine has revolutionised the recovery of such dispersed, mixed plant remains. Water is pumped through an open tank, typically converted from the ubiquitous 40-gallon oil drum. Archaeological soil is tipped in; the charred seeds and charcoal float off and are captured in a sieve, dried, and are then ready for study under a microscope. Through time-consuming comparison with modern seeds, many charred seeds can be identified to genus or species.

Most of the plant remains that we find in archaeological sites are, unsurprisingly, the result of burning of wood or dung as fuel, or are food by-products such as cereal chaff and weed seeds. Many medicinal plants are also food plants, so evidence of their presence, even as food, is still useful in showing that the plant would have been available for medical use. Evidence of actual medical use is far rarer; this is not surprising, as many medicinal plants would have been cherished and less likely to be burnt. Furthermore, many of the useful parts of medicinal plants, such as leaves or roots, are either fragile and unlikely to survive burning or, as in the case of roots, are not easy to identify.

It is a common error to assume that the presence of the seed of a medicinal plant is evidence of its medical use. If it is, say, the leaves or bark that is used, the seeds are much more likely to be evidence of the presence of the plant as a field weed. It is very rare to find definite archaeobotanical evidence of medicinal plant use.

The archaeological record for Mesopotamia is itself uneven. Archaeologists have given more attention to botanical sampling of sites dating to the beginning of farming, some 11,000 years ago, than to those dating to the cuneiform period. Later periods, such as the Greco-Roman and Islamic periods, are very poorly represented. Furthermore, excavations have often concentrated on official buildings, which tend to be swept clean, rather than on the refuse tips in which most of the plant remains reside. Nonetheless, drawing on comparisons with records from neighbouring countries, an outline picture of the presence and human connection of several hundred plants can be established.

Early village sites in Iraq fall into two groups: those predating the beginning of farming and occupied by foragers, such as Qermez Dere (10000–8800 BCE) and Nemrik (9800–8300 BCE) in northern Iraq, and early farming sites such as Jarmo (8000–7400 BCE), with evidence of crops such as einkorn and emmer wheat, barley, lentils, bitter vetch (*Vicia ervilia*) and flax. All these early sites are rich in wild fruits and nuts, which continue to be important foods for farming communities to the current day. All too few plant remains have been recovered from the initial farming settlements in lowland Mesopotamia, in the sixth and fifth millennia BCE: small deposits of charred seeds at Tell es-Sawwan (6000–5500 BCE) and Choga Mami (5300–5000 BCE) contained grains of emmer and bread wheat, naked and hulled barley, and flax, and seeds of wild prosopis (*Prosopis farcta*) and caper (*Capparis spinosa*). A similar range of crops was found at the Ubaid site of 'Oueili (4500–3500 BCE). The Uruk and Bronze Age (Early Dynastic) lowland city of Abu Salabikh (3200–2300 BCE) is one of the few sites to have been well-sampled through extensive flotation and was studied by Mike Charles; crops included einkorn, emmer and bread wheat, barley, lentil, large-seeded *Lathyrus* (possibly *L. sativus*), flax, fig, grape, and a rare find of the fragile seeds of sesame, an important summer crop. A single sample from Tell ed-Der, from stores burnt by fire c. 1630 BCE, adds date, coriander, garlic and cumin to this list. Later sites show an increasing diversity in crops; thus at the northern city of Nimrud, an Assyrian burnt level dating to the 7th century BCE includes free-threshing wheat (emmer has almost disappeared), barley, broomcorn millet (*Panicum miliaceum*), lentil, chickpea, cucumber or melon, flax, grape, fig, date, olive, pomegranate and hazelnut. Given this spotty record, the presence of a species, either at an archaeological site in Iraq, or in a neighbouring country, or in a neighbouring region, is more significant than an absence. Absence could simply reflect insufficient sampling by archaeologists or a plant that, for a variety of reasons, is unlikely to become preserved.

Traditional uses

Today, medicinal plants are used in three spheres: western biomedicine, other formal medical systems such as Unani and Ayurvedic medicine in South Asia, and what is often called folk medicine, such as that practised in families or villages. Even if a plant is used in all three contexts, its purpose will usually vary between them. Furthermore, the routes of transmission, and the degree of standardisation, are very different. Formal medical systems usually have a standard curriculum and written reference works. They change through time, but there is a shared philosophy and body of knowledge. By contrast, folk knowledge may be maintained at the level of an individual (both men and women), a household or a village. It is often passed on by oral transmission from generation to generation. And it will vary from village to village.

Herbarium specimen of red bryony (*Bryonia multiflora*). Collected near Shaqlawa, 1956, by R. Wheeler Haines

Royal Botanic Garden Edinburgh

Our knowledge of the world's folk medical traditions is very incomplete, but increasing interest from both a medical and cultural standpoint is leading to more surveys. Often this ethnobotanical fieldwork is accompanied by efforts to support the preservation of traditional knowledge among local communities. For plants used medicinally in the cuneiform period, similar or identical applications may survive to later periods, and indeed to the current day. If ethnobotanical data provide the same or similar information as described in the cuneiform tradition, the identification of the plant name is strengthened. In our research, we have drawn on ethnobotanical surveys across the region, which are very usefully summarised in a massive reference work published by Diego Rivera Núñez and co-authors in 2011, *Plants and Humans in the Near East and the Caucasus*. We have also drawn on our own experience in the countryside and markets of Iraq and its neighbours. An important document on the continuing importance of herbal medicine in today's Iraq is the work of Abdul Jaleel Ibrahim Al-Quragheely (1934–2009), who did much to make herbal medicines available in his community, and to preserve traditional uses. *A Herbal of Iraq* (2022), edited by Chris Thorogood and Shahina Ghazanfar, draws on and is inspired by his work.

Widely used medicinal plants may have many and varied uses; we have generally chosen to report those uses that are most widespread and common, as these are most likely to either represent long-standing patterns of use, or uses that have a basis in plant chemistry. When we talk of traditional uses, this is in the context of present-day southwest Asia (the Near East) unless otherwise stated.

PART 2

THE HERBAL

KEY TO THE LEVELS OF CERTAINTY	
near certain	●●●
highly probable	●●○
possible	●○○

BLACK BRYONY

Akkadian *imḫur-ašra* "It-faces-twenty"

Dioscorea communis (formerly *Tamus communis*)

●●○

Black bryony is one of the few medicinal plants for which a description in the Akkadian language is preserved. Ancient Mesopotamians did not use drawings but instead described plants with words, through comparison with other plants. They wrote about the plant *imḫur-ašra*: *"It creeps over the ground like the colocynth, its leaves look like that of henbane, its berry is red. It is called 'it-faces-twenty'."* [1] This description and the medicinal uses discussed below are so specific that they apply best to black bryony; the comparison of the alternative Akkadian name *ipšur-ašra* of the plant with the Syriac and Arabic plant names for black bryony, *fāširaštīn* and *fāširšīn* meaning "dissolving-sixty" and "dissolving-twenty", respectively, corroborates the identification.

Dioscorea communis
Reproduced with permission from *Flora of Iraq* 8: 259, pl. 66

Black bryony is a member of the yam family (Dioscoreaceae). It is a twining or climbing plant with a tuberous underground rootstock, and weak stems that climb or may creep on the ground. The leaves lack tendrils, are typically heart-shaped and are somewhat 3-lobed. The greenish-yellow small male and female flowers are borne on separate plants. The fruit is a round berry, about 12mm in diameter, orange–red when mature. All parts of the plant are poisonous, except for the young shoots, which are eaten as a green, like wild asparagus, throughout the range of the plant.

The plant is common in the lower mountains and hills of Iraq, often growing near streams or springs and in moist locations. It is distributed from western Europe to Iraq and Iran, Palestine, Lebanon and Syria and North Africa. It is absent from the archaeobotanical record, unsurprisingly given that the fruits are inedible and the plant does not grow as an agricultural weed. Only very special circumstances would lead to the archaeological preservation of the fruits.

All parts of the plant, except for young shoots, are rich in calcium oxalate crystals and chemicals that are highly irritant. These poisonous properties account for some of the traditional medicinal uses of the tuberous root, internally as a diuretic, emetic and purgative, and externally as a rubefacient; it is also used externally in Turkey for rheumatism and headaches.

The Akkadian name has many variants, one of which can be translated, namely *imḫur-ašra* "it-faces-twenty", possibly referring to the many diseases it heals; another is *ipšur-ašra* "it-loosens-twenty". The genus' former name *Tamus*, given by Linnaeus, is borrowed from a plant cited by Pliny the Elder under the name of "tamnus" (XXI, 50); *communis* refers to it being 'ordinary' or 'common'. The English name bryony has roots in the Greek *bruéin* "to sprout abundantly"; black refers to the colour of the tuberous root. With changes in understanding of plant relationships and phylogeny, *Tamus* was moved into the genus *Dioscorea* in 2002, named after the Greek physician Dioscorides. Ibn al-Bayṭār (no. 1655) gives as the Arabic name *fāširšīn*, meaning "dissolving-twenty", which may refer to the strongly laxative properties of the root.

Black bryony or it-faces-twenty could be taken as a potion, was applied as cream or poultice, and was sprinkled over affected body areas. Healers did not specify which part of the plant was used. The two principal external applications of black bryony are for paralysis or numbness and skin sores.

"It is good to remove paralysis. Dry it, crush it, rub the patient constantly with sesame oil. Grind its fresh parts, heat them in first-quality beer, prepare a poultice with sesame oil and bandage the patient." [1]

"If the patient is paralysed, the limbs are numb and he has lost weight, rub him constantly with it-faces-twenty and ghee." [2]

"It is good for long-lasting skin sores. Dry it, crush it and sprinkle it on the sore." [3]

"It-faces-twenty – drug for long-lasting skin sores – to be crushed and sprinkled over the sore." [4]

"It-faces-twenty – drug for erysipelas." [5]

The preparation of potions for stomach or intestinal disease, afflictions of the bile and jaundice suggest that black bryony was used as laxative or to induce vomiting. However, these effects are implied rather than stated.

"It-faces-twenty – drug for stomach diseases – to be crushed, drunk with first-quality beer." [6]

"If the patient's belly is aching, have him drink it-faces-twenty with beer." [7]

"If the patient's bile is sick, have him drink 2g it-faces-twenty with 80ml water." [8]

"If the patient's body is yellow, his face is yellow, he has lost weight. The name of the disease is jaundice. Crush it-faces-twenty, have the patient drink it with beer." [9]

Black bryony was also administered when a person was bewitched. Assyrian and Babylonian practitioners understood abdominal pains, marked salivation, impotence and psychological problems as possible effects of witchcraft.

"If the patient is bewitched, crush it-faces-twenty, have him drink it on an empty stomach, make him vomit." [10]

"It-faces-twenty – drug against witchcraft – to be eaten at New Moon." [11]

"It-faces-twenty – drug for a disturbed mind – to be crushed, to be rubbed constantly with sesame oil." [12]

A potion was given to women who suffered from bleeding during pregnancy.

"A woman suffers from bleeding during pregnancy, crush it-faces-twenty, have her drink it with first-quality beer." [13]

There is a debate about whether the ancient Greek name *ampelos melaina* "black vine" refers to *Dioscorea communis*, or to white bryony *Bryonia alba*. Since Dioscorides (IV, 183) describes the plant as having pale-green fruits that become black after ripening, some scholars prefer to identify it with white bryony.

Ibn al-Bayṭār (no. 1655) describes black bryony as a climbing plant that bears fruits in the form of clusters and compares the leaves with those of ivy and smilax. As for the root, it is black outside and inside the colour of boxwood.

BLACK NIGHTSHADE

Akkadian *karān šēlebi* "Fox grape"

Solanum nigrum

● ● ○

Black nightshade is another plant for which a cuneiform description is preserved. The leaves were said to look like those of the poplar, its blossom white and its fruit as black as that of camel thorn – overall an accurate description. The berries were much used in Mesopotamian medicine.

The family Solanaceae, to which the black nightshade belongs, is distributed worldwide and contains important food plants such as tomato, potato and eggplant. *Solanum nigrum* is a small annual plant with upright stems. It bears white flowers and the fruit is a spherical fleshy berry (7–8mm in diameter), dull black or purplish black when mature. It is a widespread species, found in temperate Eurasia, northern Africa and Australia, common by roadsides and waste places, in shady and moist locations. The plant is common on the alluvial plains and in the desert region of Iraq, and is occasional in the lower hills. Seeds have been found at about ten archaeological sites, mainly flanking the Mediterranean, but also in southeast Turkey. The berries are of considerable importance in traditional herbal medicine in the Near East and have been used to ease fevers, diarrhoea and eye problems.

Assyrian and Babylonian practitioners named the plant after the black berries "grape of the fox", *karān šēlebi*, whereby the reference to the fox remains unclear. Curiously enough, the Arabic name *'inab al-ṭa'lab* has exactly the same meaning as the Akkadian. The ancient Greek name, *strúchnon*, is of unknown origin and was used for other plants too. Because of the toxic properties of some members of the nightshade family, the Greek term came to be used for strychnine, a powerful poisonous alkaloid. The botanical genus name *solanum* is of doubtful etymology; the specific epithet *nigrum* refers to the black colour of the berries.

According to the cuneiform glossary of medicinal plants *Uruanna*, black nightshade was considered one of the principal plants to treat bronchial diseases, jaundice and stomach troubles, and to help during the cold winter season. *"The plant is good for alleviating very cold feeling, crush it with sesame oil or ghee, rub the patient, then he shall recover."* [1]

Other application forms are by potions, catheters and poultices. A rather specific use was made of the squeezed juice of the plant. In a rare procedure, the liquid was reduced and applied to an affliction of the eye.

"If the patient's eyes are blinking and opening wide, provoking bleary eyes, squeeze fresh fox grape and [heat] its juice in an earthenware pot. Cover the surface (of the pot) allowing on the reduced sap to dry, [mix] the dried sap with copper patina and sesame oil, [daub is eyes]"." [2]

Black nightshade could be used fresh or dried, and could be kept in private storerooms. Severe abdominal pains, digestive troubles, constipation and rectal discharge were all treated with a potion.

"If the patient suffers from severe epigastric pain, he is persistently bleeding from the nose, he is inflated and constipated, this patient suffers from 'sick buttocks' – in order to cure him: Have the patient drink fox grape with beer." [3]

"If the patient's eyes are covered with jaundice, dry fox grape and have the patient drink it with beer." [4]

As well as the potion for stomach pains, a salve could be prescribed.

"Fox grape – drug for the stomach (diseases) – to be crushed, to be heated with ghee, the patient is to be rubbed." [5]

The potion was also recommended for constriction of the urethra and calculi such as kidney stones. When used for treating gonorrhoea, the liquid was introduced with the help of a catheter.

"If a man suffers from gonorrhoea, crush roasted fox grape, fill it into his penis." [6]

"If a man has difficulty to urinate, then he suffers from constriction of the bladder: crush dried fox grape, add it to water, beer and sesame oil, let it macerate overnight under the starry sky, have him drink it on an empty stomach." [7]

Fruits of black nightshade (*Solanum nigrum*), Kashmir, India Museum (before 1879)
Economic Botany Collection, Kew 48474

Solanum nigrum
Reproduced with permission from *Flora of Pakistan* 168: fig. 1
© Pakistan Agricultural Research Council and Karachi
University, Pakistan.

"If a man suffers from a calculus, add fox grape to brewer's beer, let it macerate overnight under the starry sky, have him drink it and he shall recover." [8]

Only one case mentions the effect of the potion: when treating fever caused by heat and sun stroke the practitioners relied on the emetic effect of black nightshade.

"If a patient has fever due to heat and sun stroke, have the patient drink fresh fox grape so that he vomits, then the patient shall recover." [9]

However, more often fever was treated with poultices.

"If the patient's head holds fever, in order to remove the fever: make a poultice of fox grape and mustard water, apply it." [10]

The plant was further recommended to treat the skin of a recently born baby and a skin eruption called "stroke of the wind" or "wind blast".

"Fox grape – when the baby is born rub it constantly and as soon ..." [11]

"Fox grape – drug for the skin eruption 'wind blast' – to be crushed, rubbed with sesame oil." [12]

Applications of the potion were used in the treatment of pregnant women.

Black nightshade. Dioscorides Pedanius of Anazarbus, *De materia medica*. Constantinople, c. 512 CE. Cod. Med Gr. 1 (facs.), fol. 292v
© Österreichische Nationalbibliothek, Vienna

"If a woman is pregnant and has difficulty to give birth, crush fox grape, have her drink it with beer on an empty stomach, then she shall deliver quickly." [13]
"If a woman's bleeding cannot be stopped: crush fox grape, she shall drink it on an empty stomach, then the bleeding will be stopped." [14]

That ancient Mesopotamians were possibly aware of the pain-killing effect of the black nightshade is shown by the next case:
"Fox grape – drug for scorpion sting – to be drunk with first-quality beer." [15]

The Greek physician Dioscorides (IV, 70) reports that a plaster of the leaves, as well as their juice, was helpful for shingles, fistulas and skin ulcers, and was good for treating heatstroke in children. Al-Kindī (no. 208) records the uses of the plant in the treatment of erysipelas. Ibn al-Bayṭār (no. 1589) refers to the cooling properties of black nightshade and its uses for internal tumours of the abdominal organs, troubles of liver and spleen, jaundice, skin ulcers and burns.

CAMEL THORN

Akkadian *baltu* | Sumerian *giš-diḫ₃*
Alhagi graecorum

● ● ●

Before tilling a field in ancient Mesopotamia, there was first the hard labour of removing camel thorn and false carob (*Prosopis farcta*) from the arable land. According to administrative texts dating to the end of the 3rd millennium BCE, cutting the two spiny bushes required more manpower than ploughing and was more time-consuming than any other agricultural labour. As Al-Birūnī (973–1050 CE) writes on camel thorn in his work on simple drugs, "its growth on saline soil comes not true; it chooses the best earth, but one can see it even on the top of rocks. Should you dig it out with its root you will not stop until you reach water and moisture, and sometimes it may lead you to a depth of two hundred cubits" (Verma 1990: 91). Both the camel thorn and false carob are weedy shrubs, difficult to remove because of their long and widespread roots. A Sumerian poem praises the plant because grazing sheep would nibble at its leaves and branches. According to Akkadian plant descriptions, the blossoms of camel thorn are red; and the contours of the bush look like a small poplar.

The camel thorn is an upright, much-branched bushy perennial with thorny stems and branches. It is in the pea family (Fabaceae), with small red flowers arranged sparsely on the spine-tipped branches. The fruit is a somewhat twisted pod, irregularly constricted between the seeds. It is a widespread species from North Africa through the Arabian Peninsula and Iraq, westwards to Greece. Camel thorn is common in places where the soil is slightly saline such as in desert oases, wasteland, by fields and on the alluvial plains in the desert regions of present-day Iraq, but it is also common on sweet soils.

Herbarium specimen of *Alhagi graecorum*, collected by T.H.H. von Heldreich in Greece, 1847
Kew Herbarium K000262768

There are just a few archaeological occurrences of *Alhagi*, usually not identified to species, in Syria and Turkey. This might suggest that the process of clearing the fields described above was usually successful in removing this weed. Camel thorn is widely used in Middle Eastern traditional medicine. In Egypt it is used to treat urinary tract disorders, in Iran as a febrifuge, diuretic and thirst-quencher, and in Afghanistan to treat migraine, skin wounds and painful joints; the manna is a mild laxative.

The meanings of the Akkadian name *baltu* and the Sumerian name *giš-diḫ₃* are not known. The botanical genus is based on the Arabic word for a "camel for riding", *haǧīn*; indeed, camel thorn is a typical fodder plant for camels. The species name, *maurorum*, comes from the Latin *maurus*, itself from ancient Greek *maurós* ("dark"), referring to the black pods.

Camel thorn is believed to be one of the "manna" plants mentioned in the Bible and Qur'ān as producing "bread from heaven to eat". The manna from the camel thorn is an exudate (whether produced by insects or exuding directly from the plant is unclear), called in Persian *tarangabin* (lit. "wet honey") and appearing as a loanword in Arabic, *taranǧubīn*. It is in the form of brownish tear-shaped drops, and is an important plant product in Iran, used in traditional medicine and in the preparation of special sweetmeats.

Ancient Mesopotamian practitioners considered camel thorn to be one of the principal plants used to treat a specific disease affecting the joints and muscles of the legs called in Akkadian *maškadu* and "fever caused by sun and heatstroke": "*Drug for the maškadu joint disease – camel thorn*" [1] and "*Drug for fever because of heat and sun – camel thorn.*" [2] The use of the plant as the principal drug for "*sorrow and crying*" is obscure. [3]

Other applications include boils and ulcers of the skin, stricture of the urethra, headaches and loss of teeth. Practitioners used the root, thorn, shoot and a powder made from the dried plant. Urinary retention was treated with a potion made from the thorns.
"*Shoot of camel thorn – drug for stricture of the bladder – to be crushed and drunken with first-quality beer.*" [4]

The thorns, taken when the plant was in flower, were also used to prepare a salve, which was applied to the skin.
"*Thorn of the red flowering camel thorn – drug for boils and lumps on the skin – to be crushed in sesame oil and to be rubbed.*" [5]

Fever was treated with a salve and a small sac, filled with the leaves, that was hung around the patient's neck.

**Manna from camel thorn
(*Alhagi graecorum*),
Agra, India, 1862**
Economic Botany Collection,
Kew 59930

*"If the patient has caught fever, wrap camel thorn leaf in fleece and hang it around
the neck."* [6]
*"Shoot of camel thorn – drug for the fever demon Lamashtu – to be crushed and rubbed
with sesame oil."* [7]
    The root was employed to avoid loss of teeth and to cure headaches.
*"Root of camel thorn which has been dug out without seeing the sun – drug for loss of
teeth – to be dried, crushed, mixed with sesame oil and applied to the teeth."* [8]
*"If the 'spirit-of-the-dead' has caught a patient and he suffers from migraine attacks,
look for camel thorn root and pull it out, ply it together with red wool and bind it around
the patient's head."* [9]
*"If severe headaches have caught the patient, take camel thorn that rustles in the wind,
wrap it into fleece, fix it with red wool and bind it around his front, then the patient
will get well."* [10]
    Assyrian and Babylonian practitioners ascribed to the plant an intoxicating
or narcotic effect as stated in the following recipe:
*"In order to intoxicate a person, wrap powdered camel thorn in a linen, steep it in
wine, have the person drink the wine and then he shall be stupefied."* [11]

    Camel thorn is missing from the *De Materia Medica* of Dioscorides. Ibn al-
Bayṭār refers several times to camel thorn and manna. The leaves were used in eye
drops and, instilled into the nose, to treat chronic headaches (Ibn al-Bayṭār no. 553).
The manna was effective for chest and belly troubles, stomach ache, febrile cough,
acute fever, liver inflammations and to evacuate bile (Ibn al-Bayṭār nos. 408, 1380).

CEDAR

Akkadian *erēnu* | Sumerian *giš-erin*
Cedrus libani

● ● ●

The genus *Cedrus* has only three species, and is distributed in northwest Africa, the eastern Mediterranean and the Himalayas. The eastern Mediterranean cedar is *Cedrus libani*, commonly known as the cedar of Lebanon. It is native to a region spanning Turkey and Lebanon, and in Cyprus. It is a large tree that can grow to a height of 40m (130ft), with a trunk girth of 8m or more. The almost horizontal branches of this cedar can have a spread of up to 30m, making it the most magnificent and stately tree in the Middle East. As in most non-flowering seed plants, the male and female cones are separate. The seed-bearing (female) cones are woody at maturity, barrel-shaped, resinous, and borne upright at the end of branches. Cedars and some species of pines contain resinous ducts in their wood, giving resistance to insects and pathogens, thus making the wood scented, and more durable and lasting.

The long trunks and the durability of the wood made cedar one of the preferred trees for the construction of temples and palaces. However, it did not grow in Mesopotamia proper and had to be imported from other countries. The principal sources of cedar wood were the Nur or Amanus mountain range that stretches from southern Turkey to Syria, and the extensive Mount Lebanon range that runs parallel to the Levantine coastline. Throughout Mesopotamian history, kings praised themselves for bringing timber from these "cedar mountains" or receiving as tribute cedar and other trees from their vassals. Sometimes the tribute also included cedar tar, which is obtained by destructive distillation (pyrolysis) of the tree stumps and the branches. Some of the first-millennium BCE Assyrian kings, such as Sargon II (721–705 BCE), had their palace reliefs decorated with scenes showing how the felled tree trunks were tied into rafts and towed by boats.

Cedar wood is found sporadically in Near Eastern excavations, usually in circumstances reflecting its high value and aromatic properties, as in the coffin of Tutankhamun in ancient Egypt (c. 1320 BCE) and as a component of furniture from the royal tombs at Gordion, Turkey (c. 700 BCE). Today cedar is little used in traditional medicine. Cedar tar is sold by herbalists in southeast Turkey; in Lebanon, a decoction of twigs, or cedar oil mixed with olive oil, is used externally to treat rheumatoid arthritis.

Assyrian soldiers rafting timber. Wall relief at the Palace of Nineveh, c. 700 BCE

M.P.E. Botta & M.E. Flandin, Monument de Ninive I. Paris: Imprimerie nationale, 1849, pl. 41

Oil of cedar of Lebanon (*Cedrus libani*), 19ᵗʰ century

Economic Botany Collection, Kew 76465

The meaning of the Sumerian and Akkadian terms for cedar is not known. The English name draws on the Latin *cedrus* and Greek *kédros*, which are related to the Indo-European root **ked-* "to smoke, to burn incense". The Arabic name *arz* and Hebrew *'erez* are linked to the Semitic root meaning "to be firm", which alludes to the massive columnar trunk of the cedar.

The wood was used for roof beams, doors and columns and to manufacture cultural and ritual objects. Because of its sweet, warm, balsamic aroma, the wood was made into incense. Tales were woven around the cedar growing far beyond the borders of the known world – according to a singular cuneiform clay tablet, the so-called Babylonian map of the world that shows a bird's-eye view of the surface of the earth. One of these tales entered the famous epic about the legendary king Gilgamesh. It narrates how the king and his soul mate Enkidu embark on an adventurous trip to the cedar forest to face its guardian, the monstrous Humbaba, and eventually kill him. Uniquely in Mesopotamian literature, the forest is described as a noisy place full of the penetrating buzz of cicadas, roaring monkeys and chirping birds. In the mythological view of Mesopotamians, the cedar forest marked the Western Gate of the world where the

Cedrus libani

Reproduced with permission from A. Farjon,
A Handbook of the World's Conifers

sun would set and the sun god would enter the underworld. There he would spend the night in order to rise the next morning in the east, stepping out through the Eastern Gate that was located in a mountain range on the eastern edge of the Persian Gulf. This worldview explains why cedar wood was used in rituals addressing the sun god, and why cedar tar is especially employed for diseases that are associated with the "spirits-of-the-dead" who are living in the underworld.

Assyrian and Babylonian healers used cedar oil, charcoal, resin and tar. The oil was made on a base of sesame oil. The resin was rarely employed; cedar tar appears far more often and was called "blood of the cedar" (*dām erēni*). The oil was used as salve to treat skin eruptions and muscular troubles. Resin and tar were prescribed in case of ear infections. Cedar charcoal served to alleviate pains that could occur during pregnancy.

"If the patient suffers from deep-red eruptions on the skin, rub him constantly with cedar oil." [1]

"If the patient's head is covered with sweet-smelling scab, shave his head, rub him with cedar oil and crush cedar in order to sprinkle it on the sore, then the patient shall get well." [2]

"If the patient's body ... is afflicted with ..., paralysis and numbness and he has lost weight, rub the patient with cedar oil." [3]

"Make a pad for the ears with cedar resin and cedar tar, place it into the ears, then the patient shall recover." [4]

"If a woman suffers from problems during pregnancy, crush cedar charcoal, wrap it into fleece." [5]

Cedar tar was used as a carrier for salves that practitioners recommended to treat afflictions that they attributed to the attacks of the "spirit-of-the-dead". If the "hand" of such a spirit continuously touched the patient and could not be removed, the practitioner did the following:

"Roast a mineral, coral and false carob seed over coals, crush them, mix them with cedar tar, recite two incantations seven times over the salve, rub the patient wherever it hurts, then the pains will be alleviated." [6]

If the "spirit-of-the-dead" attacked the temples and the eyes, cedar tar mixed with other ingredients would be used as an eye daub. If there was a roaring in the ears – the spirit was believed to enter the body through the ears – cedar tar would be used as lubricant for an ear pad or mixed with other ingredients directly instilled in the ear.

"If pus discharges from the patient's ear, mix blood from the kidney of an ox with cedar tar, trip it into the patient's ear." [7]

In Mesopotamia the diviner was one of the religious experts who made a special use of cedar. His duty was to inspect the entrails of the sacrificial lamb in order to make predictions for the future; the profession was called in Akkadian *bārû* literally "he who examines", which accurately describes his task. Before the actual inspection took place, the diviner had to chew cedar wood, place cedar wood in the mouth of the sacrificial sheep and burn cedar wood as incense.

Cedar trees. Dioscorides Pedanius of Anazarbus, *De materia medica* fol. 251v, Constantinople, mid-10th century

© The Morgan Library & Museum, MS M.652, purchased by J.P. Morgan

The ceremony started with a prayer, *"I invite the gods by means of cedar, let resin and cedar wood bring you forth!"* (Starr 1983: 37), to summon the sun god and the weather god, both patrons of divination. The sweet and warm smell of the wood attracted the two gods. An additional benefit was perhaps a soothing and relaxing effect on the diviner's mind that prepared him for the strenuous work that lay ahead.

Dioscorides (I, 77) praised the properties of cedar wood as driving away insects and worms. He recommended cedar for the treatment of eyes, ringing ears, toothache, sore throats and skin lesions caused by insect and snakebites. As an enema, it had abortifacient and worm-expelling properties. The cedar from Syria drew the menses and was helpful for coughs and spasms. Al-Kindī (no. 234) refers in particular to cedar tar, which that was supposed to be the best medicine for treating the conditions of toothache, ear infections and insanity, and as antidote for poisons.

COLOCYNTH, BITTER APPLE, BITTER GOURD

Akkadian *irrû* "Tangling (plant)" | Sumerian *ukuš₂-ḫab* "Bitter melon"
Citrullus colocynthis

● ● ○

The cuneiform sources are highly informative about the many medicinal uses of the colocynth. Medical recipes refer to the seed, apple, pulp, root, leaf and shoot of the colocynth. The plant did not belong to the usual spices used in cuisine, possibly because it was a desert plant and as such not as easily available as garden produce, but maybe also because of its potential toxicity. Occasionally, however, the bitter and pungent taste of colocynth gave meat dishes a special flavour. Practitioners described the plant, which was known as "bitter tasting melon", with the following words: *"Its tendrils creep over the ground like those of the (normal) melon, its leaves are divided and its flower is yellow."* [1]

The colocynth is a common desert creeper found from North Africa, the Arabian Peninsula and Mediterranean Europe to South Asia. It is a perennial, producing annual stems that trail on the ground from a woody base. The leaves

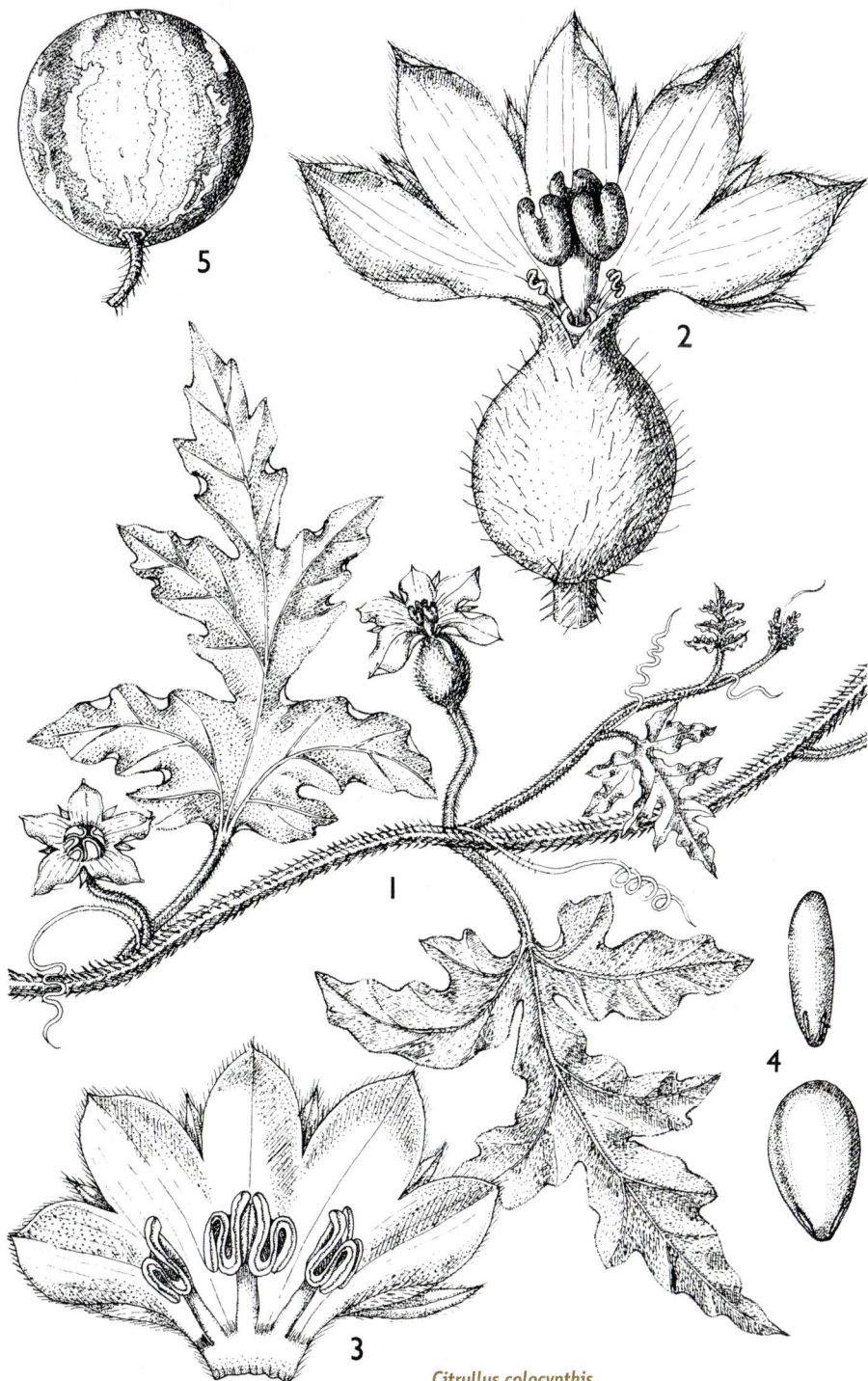

Citrullus colocynthis
Reproduced with permission from *Flora of Iraq* 4(1): 195, pl. 37

Fruit of colocynth (*Citrullus colocynthis*), Iraq, 1850s
Economic Botany Collection, Kew 54492

are deeply lobed, with tendrils opposite the leaves. The small yellow flowers produce spherical fruits 4–10cm in diameter. These are yellow at maturity, with seeds embedded in a bitter dry spongy pulp. The fruit stays intact for several months, eventually drying and breaking up irregularly. The fruits contain cucurbitacins, chemicals that are very bitter and with strong purgative properties, which explains its traditional use as a laxative.

The seeds have been found at a few archaeological sites in different parts of southwest Asia. Colocynth is in the same botanical genus as watermelon (*C. lanatus*), of which seeds were found in the tomb of Tutankhamun (c. 1320 BCE). There is no archaeological evidence for the presence of watermelon in ancient Mesopotamia. The flesh of the colocynth is a purgative and is widely used in Near Eastern traditional medicine as a laxative and for other stomach problems, including intestinal parasites. In Oman, a poultice made from crushed leaves and garlic has been used to relieve pain and itch from bites and stings, and crushed fruit mixed with oil is used to relieve pain in joints.

The Akkadian name *irrû* means literally "the tangling one", alluding to the vine-like stems. Sumerian *ukuš₂-ḫab* is a compound name. The term *ukuš₂* refers to the melon and was also used as a generic term for climbing plants with round or gourd-shaped fruits; the word *ḫab* means "bitter". The genus name *Citrullus* refers to the citrus-like shape of the fruit. The name is derived from *citrus*, probably a corruption of *cedrus*, and the diminutive suffix *–ullus*, a corruption of *unulus*, the diminutive of *unus* meaning "small size", hence "small citrus". The term came to be used for cucumbers as well, probably because both fruits were pickled in brine for preservation. *Colocynthis* comes from the Greek *kolokúnthē*, which is of unknown origin. The English name colocynth goes back to the Latin form and the other English names refer to the bitter taste and shape of the fruit. The Arabic name is *ḥanẓal* or *ḥandal*. The 12th-century Sephardic philosopher Maimonides called the colocynth "bitter of the desert" (*murrār aṣ-ṣaḥrā'*).

Babylonian and Assyrian healers used colocynth externally for the treatment of skin lesions and haemorrhoids, and internally as a strong purgative and diuretic and for the treatment of pregnant women. The skin afflictions range from sweet-smelling scab on the head to gangrene of the feet and fissures on the heels. Leaf, root, pulp and the flour of the dried fruit or leaves were crushed and usually mixed with suet to prepare a thick skin cream. After application, the skin sore could be covered with dressings or bandages. Occasionally, the powder of the dried plant parts was sprinkled on the wound.

"If the patient's head is full of sweet scab, shave the head, roast colocynth root, crush it, sprinkle it over skin sores." [2]

"If the patient suffers from gangrene of the feet: rub him with colocynth flour mixed with suet." [3]

"If the patient suffers from gangrene of his feet: crush dried colocynth leaf, rub the surface of the sore with the leaves in the acute stage of the illness or sprinkle the sores with the leaves." [4]

"If a blister appears on the patient's foot, ripens like an abscess that exudes, he shall recover. Crush dried colocynth, sprinkle it over the sore." [5]

"If the patient's heel is fissured, crush dried colocynth, mix it with suet, bandage the foot." [6]

"If external haemorrhoids are bleeding too much, remove them with a knife, roast colocynth shoot, mix it with suet, place it repeatedly on the man's anus, then he shall recover." [7]

Urinary stricture was treated with a potion made of the seed.

"If a man suffers from stricture of the bladder and retains urine, crush colocynth seed, he drinks it with beer." [8]

"If a man suffers from calculi such as kidney stones, he drinks colocynth with beer and shall recover." [9]

The purgative effect was used to treat a variety of severe digestive problems. An uncommon recipe recommends the intake of pills made of the pulp. Practitioners administered the diuretic medicine mixed with beer, but for digestive disorders, they often prescribed sweet ingredients such as date syrup, honey or date beer, to distract from the extremely bitter taste of colocynth.

"If an infection of the bile has stricken the patient: crush colocynth shoot, have him drink it with wine, syrup and filtered sesame oil, the patient shall vomit and then recover." [10]

"If the patient's bowels are repeatedly inflamed (and) the feet are all the time swollen: crush dried colocynth apple, mix it into a dough with roasted grain, make seven and seven pills. Coat them with wild honey, the patient swallows them and drinks date beer. Then he will purge through the anus and recover." [11]

"If the patient's belly is sick, crush dried colocynth leaf, sift it, add it to a mixture of syrup, first-quality beer and filtered sesame oil, have him drink it on an empty stomach, then he shall recover." [12]

Occasionally, practitioners chose an external treatment for constipation and digestive problems.

"Colocynth. It is good for constipation. Dry it, crush it and mix it with sesame oil. Apply it repeatedly to the anus, and the patient will recover." [1]

"If the patient's bowels are inflated: place colocynth pulp on the anus." [13]

"If the patient suffers from illnesses of the bile: crush dried colocynth, sift it, mix it with finely ground flour, prepare a poultice with mustard water, spread it on a leather strip, bandage the patient." [14]

In order to get rid of symptomatic excrements, a potion on the basis of beer was prescribed.

"If a man defecates and evacuates from his anus either bloody excrement, pus or mucus; in order to heal him: [he shall drink] colocynth pulp [with beer]." [15]

Colocynth. Dioscorides Pedanius of Anazarbus, *De materia medica*. Constantinople, c. 512 CE. Cod. Med Gr. 1 (facs.), fol. 190v
© Österreichische Nationalbibliothek, Vienna

It is not clear how colocynth pulp was administered to a pregnant woman showing signs of leaking amniotic fluid.

"If the amniotic fluid of a pregnant woman passes out and there is danger to lose the foetus: take 8.33g of colocynth pulp ... Then the amniotic fluid shall be held and the foetus will be kept." [16]

Dioscorides (IV, 176) mentions the purging properties of the pulp when taken as a pill. Dried and ground up it was used for enemas to treat hip ailments, paralysis and colic; the plant removed phlegm and bile and served as an abortifacient. A rinse was used to alleviate toothaches. Colocynth suppositories were helpful as a laxative. Al-Kindī (no. 84) recommends colocynth for treating itching skin and rheumatism, and to excrete phlegm. Al-Dīnawarī describes in detail how the colocynth was harvested, the seeds extracted and washed, and the medicine prepared (*The Book of Plants*, 287).

CORIANDER

Akkadian *kisibirru, kusibirru* | Sumerian *še-lu*$_2$

Coriandrum sativum

●●○

Coriander was an important kitchen herb in ancient Mesopotamian cuisine, with some medicinal uses. The plants are slender, 20–80cm tall, with a characteristic sweet-musky smell. As with other members of the carrot plant family (Apiaceae), the white or pinkish flowers are clustered in umbels. In each umbel, the flower stalks all meet at a common point, like umbrella ribs. Coriander is widely cultivated as a garden herb on the plains of central and southern Iraq, and probably also in the lower mountain valleys in the north. It is also a weed in gardens and fields but does not grow truly wild in Iraq. Coriander was probably first taken into cultivation in the Near East, but its exact wild ancestor and, thus, area of origin are not known.

Although probably most familiar to readers as a fresh herb, now widely available in supermarkets, the fruits were in the past perhaps the most important product because they were the most portable. Unlike fresh leaves

Coriandrum sativum
Reproduced with permission from *Flora of Iraq* 5(2): 251, f. 93

of coriander, which do not dry well, the fruits can be stored and easily traded. Leaves and fruits of coriander are used to flavour soups, stews, curries, salads etc. in many countries, and in the Hadramaut in southern Arabia, they are used to flavour bread.

The young stems are sometimes used as a spice in Georgia while the dried stems provide fuel. The fruit contains up to 1% essential oil, of which the main component is coriandrol, which can be extracted by soaking the fruit in water for about 14 hours and then by distillation. The essential oils of coriander stimulate gastric secretion and thus quicken the appetite and enhance digestion. This explains the importance of the spice in cuisine and its common use in traditional medicine as a carminative to relieve flatulence or abdominal pain. The oil is also used in the manufacture of perfumes, soap and liqueurs, and to improve the taste of medicines; the fruit is also sometimes employed as a flavouring and spice in pastries and even in canned meats. The fruits have a wide range of uses in traditional medicine throughout the Near East, most often for the digestive system; other applications include use as an aphrodisiac and, when smoked, to alleviate toothache.

Although the oldest coriander fruits found so far date to about 6000 BCE, and were found at the cave of Nahal Hemar (Israel), finds at later sites are scattered and infrequent, including a single fruit in the Parthian levels (c. 2,000 years ago) at Larsa, southern Iraq. However, 62 fruits were found on a floor at Tell ed-Der (part of ancient Sippar), eastern Iraq, dating to about 1600 BCE. By

Fruits of coriander (*Coriandrum sativum*), India, 1924
Economic Botany Collection, Kew 56052

the middle of the 3rd millennium BCE, coriander was mentioned in cuneiform administrative records. It was usually cultivated together with onions, pulses and flax in large garden plots. The large amounts of the harvested herb – given in measures of capacity – between 90 and 140l per plot provide a picture of its importance as a condiment in Mesopotamian cuisine. According to cuneiform culinary recipes, it was added to enhance the flavour of meat and vegetable dishes; together with leek, garlic and cumin, it formed one of the standard spice mixtures for marinating meat.

The Akkadian word *kisibirru* or *kusibirru* for coriander is a loanword from an unknown language; the Arabic and Sanskrit terms *kuzbara* and *kustumbarī*, respectively, seem to go back to the same word in both cases very probably borrowed from Akkadian. The English name is derived from Latin *Coriandrum*, itself derived from the ancient Greek *kóris* or *koríannon*, the name for a bug. It was apparently the strong "buggy" smell of the leaves that made ancient Greeks choose this name.

Above all, Babylonian and Assyrian practitioners treated worms and other parasite infestations with coriander. We do not know whether they used the whole plant, or just a part such as the fruits, because of the highly abridged style of their recipes.

"If a patient is sick with worms, crush coriander, have him drink it with syrup and then they shall be expelled." [1]

This recipe recommends taking coriander with syrup, possibly made from dates. Other remedies for treating worms refer to water, vinegar and the ubiquitous beer. The sweetish liquid was no doubt to mask the soapy taste of coriander. A very similar recommendation is given by Dioscorides (III, 63); he suggested taking the fruits with syrup made from grapes, which had the same economic and cultural significance in ancient Greece as dates in ancient Mesopotamia.

The other medicinal reference to coriander is in the form of a precaution for patients with eye troubles.

"Coriander – a patient whose eyes are affected should not eat it." [2]

Al-Kindī (no. 263) refers to the use of coriander for headache. Ibn al-Bayṭār (no. 1926) mentions that the leaves were used as poultice for eye troubles and for treating headaches.

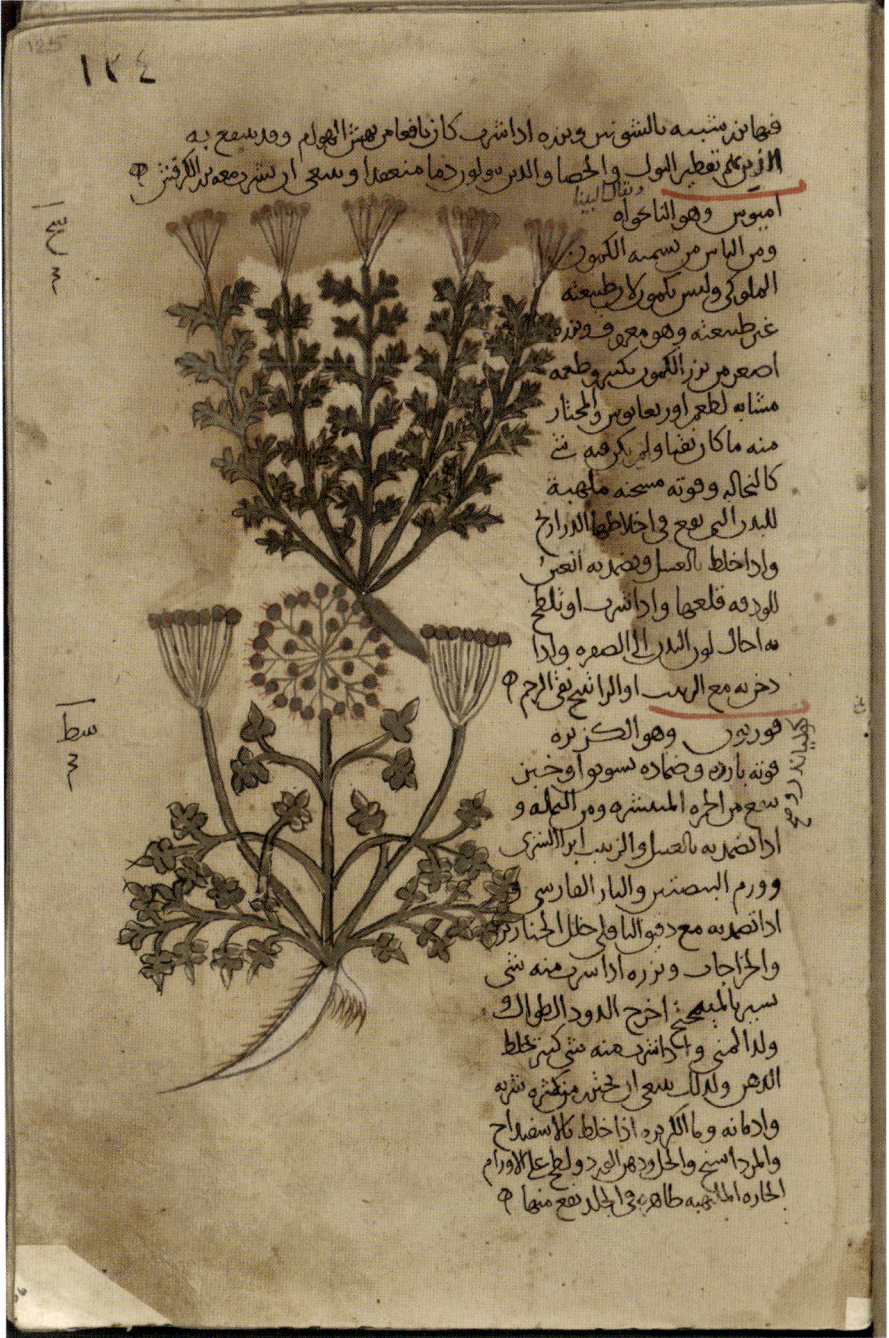

Coriander. *Kitāb al-Ḥashāʾish fī Hayūlā al-ʿIlāğ al-Ṭibbī*. Arabic translation of Dioscorides' *De materia medica*; revised edition of Abū ʿAbdallāh al-Nātilī (380/999) of the original Arabic translation by Ḥunayn ben Isḥāq (d. 260/873), Leiden, Universiteitsbibliotheek, MS Or. 289, fol. 125r

CUMIN

Akkadian *kamūnu* | Sumerian *gamun*

Cuminum cyminum

● ● ○

Like coriander, cumin was one of the crops grown in ancient Mesopotamia, at least from the 3rd millennium BCE. It was cultivated in huge garden plots together with legumes, onions, garlic, lettuce and coriander. The fruits were commonly used in cuisine to flavour meat and vegetable dishes. To enhance their fragrance, they were often roasted before being added to stews, soups or meat marinades. Cumin was so indispensable as a kitchen spice that it formed part of the provisions taken on journeys. Writing in Roman times, Pliny the Elder stated that cumin was the best of all herbs for squeamish and delicate stomachs (XIX, 8).

Cumin belongs to the carrot family (Apiaceae), which contains many foods (carrot, parsnip, celery), spices (anise, asafoetida, caraway, coriander, cumin, dill, fennel) and some genera containing lethal alkaloid poisons (hemlock, *Conium*; water hemlock, *Cicuta*). Cumin is a slender annual or biennial herb with leaves divided into narrow segments. The minute deep pink or white flowers are arranged in umbrella-like heads and the fruit is oblong to ovoid with five prominent ridges along its length. The original distribution of cumin extends from the Mediterranean to Central Asia, but it is widely cultivated especially in India, China and North Africa. As is often the case with spices, the fruits have only been found at a handful of sites, in this case dating to the Bronze Age and Iron Age, including 145 fruits on the floor of Tell ed-Der (ancient Sippar), dating to 1600 BCE.

Fruits of cumin (*Cuminum cyminum*),
Baghdad, undated
Economic Botany Collection, Kew 92899

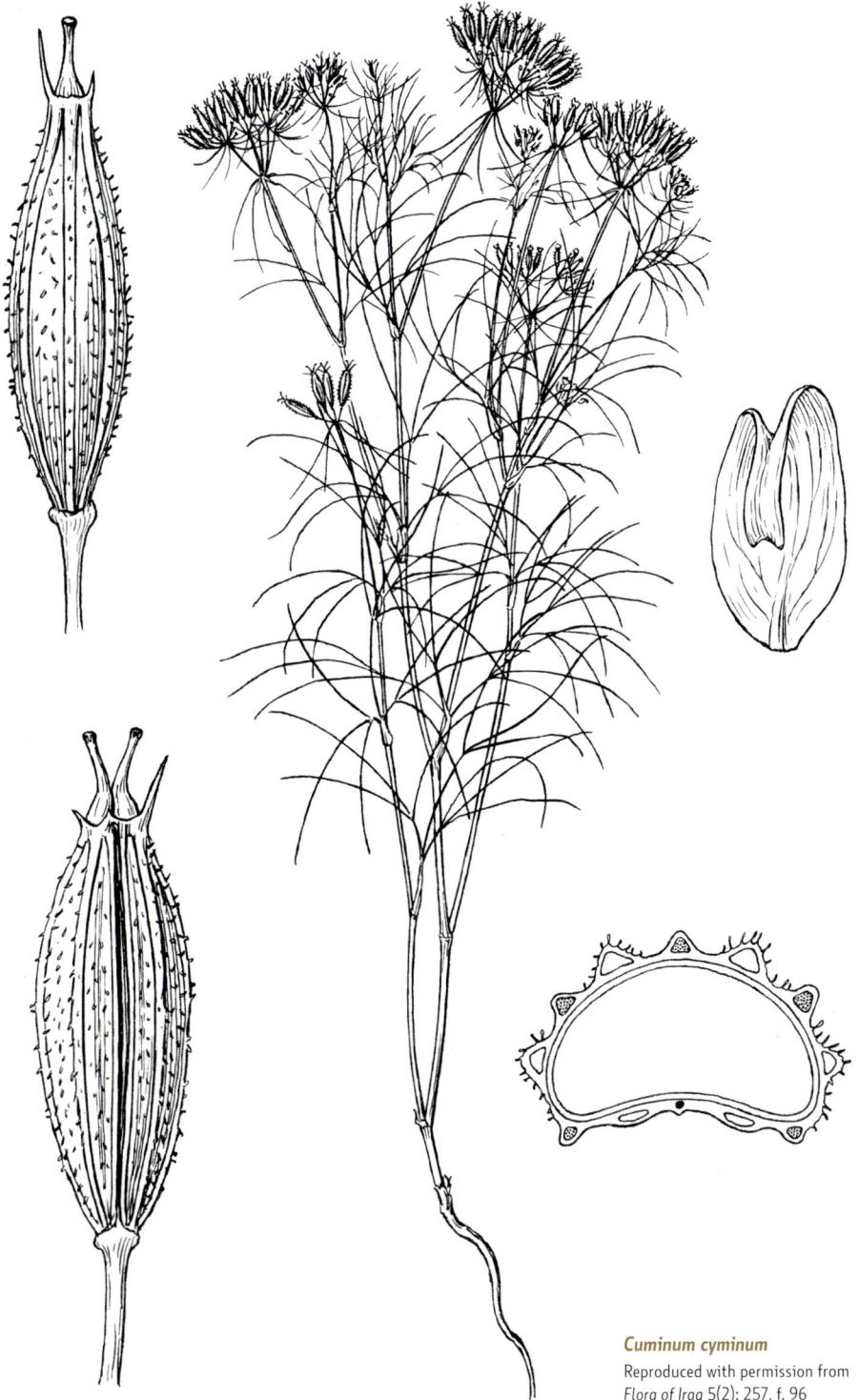

Cuminum cyminum
Reproduced with permission from
Flora of Iraq 5(2): 257, f. 96

The fruits contain essential oils, principally cuminol, that give cumin its unique aroma and flavour when crushed and used in food, and which have medicinal properties. In modern traditional medicine in the Near East, cumin is valued for digestive conditions, as a carminative, diuretic and stomachic. It was used to treat hoarseness, dyspepsia and diarrhoea, and was taken to relieve pains of childbirth.

The Akkadian name for cumin, *kamūnu*, is a wanderword that shows up in numerous languages and geographical regions. From Semitic Mesopotamia, it entered the Sumerian vocabulary as *gamun*; the ancient Greeks knew it as *kúminon*, in Arabic it is called *kammūn* and in English cumin.

The fruits were usually crushed before their use, and depending on how they were administered, mixed with oil, beer, tree resin or lard. They were available in Assyrian storehouses or could be ordered directly from gardeners, as shown by cuneiform tablets from the middle of the 2nd millennium BCE. As a single ingredient, the fruits were used to treat skin lesions, scorpion stings, inflammations and fever.

"Drug cumin – drug for inflamed buttocks – to be crushed and anointed with oil." [1]

"Drug cumin – drug for a red skin lesion – to be crushed and anointed with oil." [2]

"If a person is stung by a scorpion, crush cumin, have the patient drink it with beer and eat it; then he shall recover." [3]

"If a person has fever, crush cumin, add it to oil, annoint the patient, then he shall recover." [4]

Cumin. Dioscorides Pedanius of Anazarbus, *De materia medica* **fol. 80r, Constantinople, mid-10th century**

© The Morgan Library & Museum, MS M.652, purchased by J.P. Morgan

Assyrian and Babylonian practitioners included cumin in some compound medicines. To heal pus discharging from the ear, cumin was mixed with tree resins and lard, and then applied as a plug to the ear. The ear, however, was first treated with pomegranate juice.

"If pus discharges from a patient's ears, drip pomegranate juice into the ears, mix cumin and resin with lard, wrap the mixture into fleece, introduce it to the patient's ears." [5]

Dioscorides (III, 59) recommended cumin for colic and abdominal distension, to treat wild animal bites and difficulties with breathing, and to bring relief from testicular inflammations and inflammations due to stones. He also mentions that it kept menstruation and nosebleeds in check, and smeared on the skin it makes it paler. Al-Kindī (no. 266) refers to the use of cumin in an oil to treat rheumatism in the knees and other joints and as stomachic. Ibn al-Bayṭār (no. 1967) adds that a mixture of oil and cumin taken as suppository stops excessive menstruation.

DATE PALM

Akkadian *gišimmaru* I Sumerian *gišimar*

Dried date fruit I Akkadian *suluppu* I Sumerian *zulum*

Fresh date fruit I Akkadian *uḫinnu* I Sumerian *uḫin*

Phoenix dactylifera

● ● ●

It is hard to imagine the Mesopotamian landscape without date palms. Herodotus (484–426 CE), in describing Babylonia, wrote: "Palm trees grow in great numbers over the whole of the flat country, mostly of the kind which bears fruit, and this fruit supplies them with bread, wine and honey" (Book I, 193). Gardens, especially in southern Mesopotamia, were essentially plantations of date palms. The shade of the tree creates a microclimate that is cooler, and with a higher average humidity, which is the reason why these gardens were organised with several vegetation layers. Date palms occupied the upper layer, fruit trees such as pomegranates, figs or trained grapes vines the middle layer, and plots of cereals, legumes and vegetables such as

Sculptured relief panel from Ashurnaṣirpal II's Northwest Palace at Nimrud (ancient Kalḫu), showing a supernatural figure flanking a sacred tree. He is holding in his left a bucket and in his right hand a cone whose exact nature is not clear. It has been suggested that the cone resembles the male date spathe, used to artificially fertilise female date palms, c. 883–859 BCE

© The Metropolitan Museum of Art, New York, 31.172.1, Gift of John D. Rockefeller Jr., 1931

onions, garlic and leeks the lowest. Because the date palm stabilises the soil, Mesopotamian farmers grew it along rivers and canals.

Phoenix dactylifera, the date palm, is a dioecious plant (with the male and female flowers borne on separate trees), belonging to the palm family (Arecaceae). It produces basal suckers that are used in propagation, rather than seed. This ensures the descendant trees are genetically identical and bear the same quality of fruit. Wild populations are pollinated by wind and produce an equal proportion of male and female individuals. By contrast, in cultivated date gardens, one male to 25–50 female trees are planted in a date grove, and pollination is carried out manually by cutting off male flowers and dusting their pollen on female flowers. The artificial pollination of date palms is reported in cuneiform documents from the reign of the Babylonian king Hammurabi

(1792–1750 BCE), possibly reflecting a practice carried out since at least the early 3rd millennium BCE. Four of Hammurabi's Laws govern the cultivation and pollination of date palms within orchards. The Akkadian verb that describes the process of pollination is "to ride, to mount", which is also used in the context of human and animal copulation. Later, in the 1st millennium BCE, Assyrian winged guardian gods would be depicted on palace reliefs holding a cone that was used for the pollination; the relief panel, illustrating a stylised tree, typical in Assyrian art, symbolises fertility.

Archaeobotanical evidence shows that date palms probably grew wild in several parts of the Near East, but these original populations have mostly long been displaced by agriculture. Genetic evidence suggests that wild date palms still survive in Oman, and that the Arabian Peninsula was the region where date palms were first taken into cultivation. The earliest finds of date stones from Arabia are dated about 5000 BCE and from Mesopotamia 4700–4000 BCE. After 3000 BCE, larger-seeded (and presumably larger-fruited) forms appear. Date stones are robust, even when burnt, and have been found at over 50 sites in the Near East. Iraqi finds include abundant burnt seeds in tomb offerings from Ur, dated to 3000–2000 years BCE.

The traditional medicinal uses of date fruits in the Near East are so diverse that it is hard to find a common theme, but they include illnesses related to eyesight, the stomach, throat and chest infections, and the skin (the latter treated externally)

The fruit is highly nutritious and an important source of energy; moreover dates could be easily preserved, stored and transported. Dates were eaten fresh (called in Sumerian *uhin*, Akkadian *uhinnu*) or dried (Sumerian *zulum*, Akkadian *suluppu*). They were used to prepare syrups (Sumerian *lal₃*, Akkadian *dišpu*), date beer (Sumerian *kaš zuluma*) and date wine (Akkadian *kurunnu*). The wood was used for furniture, the fibres to manufacture ropes, the fronds for constructing houses, brooms and baskets, and the mid-ribs of the fronds for making beds and mats. The importance of the date palm for the Southern Mesopotamian economy explains why the Assyrian army systematically used the destruction of date palm groves as a military strategy to threaten the Babylonians – be it to force a city to capitulate or to punish rebel kings.

The meanings of the Sumerian and Akkadian names for date palm, *gišimar* and *ǧišimmaru*, respectively, are unknown. The origin of the botanical name *Phoenix* is obscure. It might relate to the Phoenicians or to the mythical bird of Arabia that flew to Egypt every 500 years to be reborn. The English term for

Assyrian soldiers felling date palms during the despoliation of a city in southern Mesopotamia. A.H. Layard. *Discoveries Among the Ruins of Nineveh and Babylon.* New York: Harper & Brothers, 1871, p. 500

the fruit of the date palm is derived from the Latin *dactylus* and Greek *daktylos* "finger". The species name *dactylifera* comes from *dactylus* "finger" + *fer* "bearing", referring to the oblong fruits, the dates.

The Arabic vocabulary for the date palm and its products is as large as the cuneiform lexicon – evidence that highlights its place as a food source in ancient Arabia. Its economic importance today is reflected in the names given to each part of the date palm: *naḥl* refers to the tree, *ǧummār* to the date heart, *busr* and *balaḥ* are names for the unripe date, *raṭab* the ripe, *qasab* is the dried date, and *nawāh* refers to the date pit (stone).

Assyrian and Babylonian practitioners used above all the by-products of the date, namely the stones, the fruit peel and the leaf frond. The consumption of the fruit itself was suggested in rather exceptional circumstances. If a person was haunted by the 'spirit-of-the-dead' so that the left ear would ring, dates should be eaten. The recommendation is a good example of the confluence of medical and magical thought. In the Mesopotamian mind, pain was understood as a negative sign, but since it appeared on the left side, which is related to evil, the overall significance was positive following the rule 'minus times minus equals a plus'.

"If a man's left ear is constantly ringing, he will make profit – he should eat during seven days dates and porridge, then he shall get well." [1]

Roasted and crushed date pits were used to treat sores of the skin including foot ulcers, gangrene, and sweet-smelling wounds on the head. The pits were mixed with a greasy carrier and daubed on the affected part and applied to the skin with a bandage.

"If a patient suffers from an ulcer on the foot, roast date pits, crush them, mix them with sesame oil into a lotion, daub the foot and the patient shall recover." [2]

"If a patient suffers from gangrene of the feet, crush date pits, mix them with lard and daub the feet." [3]

"If the patient's head is covered with sweet-smelling sores, roast date pits, crush them, apply them with a bandage, then the patient will get well." [4]

Date pits were also taken internally, as a potion for jaundice, a pill to bring relief from dry eyes, and as a suppository or wad of absorbent material for bleeding.

"If a patient's eyes are filled with jaundice, crush date pits, let them macerate in beer overnight under the starry sky, have him drink it the next morning." [5]

"If a patient suffers from dry eyes, roast ground date pits, crush them, knead them with mustard water, form a pill, he shall swallow it before eating." [6]

"If a man loses repeatedly blood with his stool like a woman who suffers from vaginal bleeding or spotting during pregnancy, examine him first and if it is not a disease affecting the anal area, he suffers from a bowel disease: wrap a date pit in red wool, sprinkle it with cedar oil, introduce it into his anus." [7]

"If a woman suffers from vaginal bleeding or spotting during pregnancy, roast date pits, crush them, wrap them into fleece, introduce it into her vagina." [8]

The frond rib was used to treat migraine.

"Frond rib of a date palm that was not exposed to wind – drug for migraine." [9]

Another by-product was the date peel, which was applied as a suppository in case of a severe gastrointestinal disease.

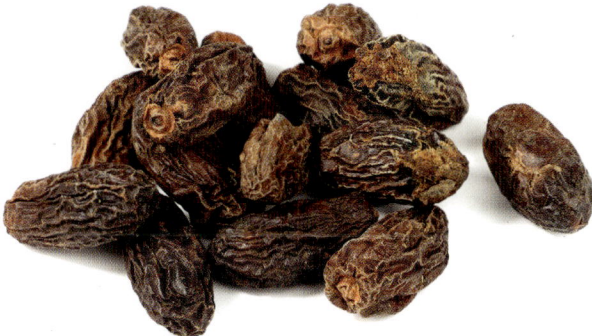

Fruits of date palm (*Phoenix dactylifera*), Busra, Iraq, 1887
Economic Botany Collection, Kew 35985

Date palm. *Kitāb al-Ḥashā'ish fī Hayūlā al-'Ilāğ al-Ṭibbī*. Arabic translation of Dioscorides' *De materia medica*; revised edition of Abū 'Abdallāh al-Nātilī (380/999) of the original Arabic translation by Ḥunayn ben Isḥāq (d. 260/873), Leiden, Universiteitsbibliotheek, MS Or. 289, fol. 44a

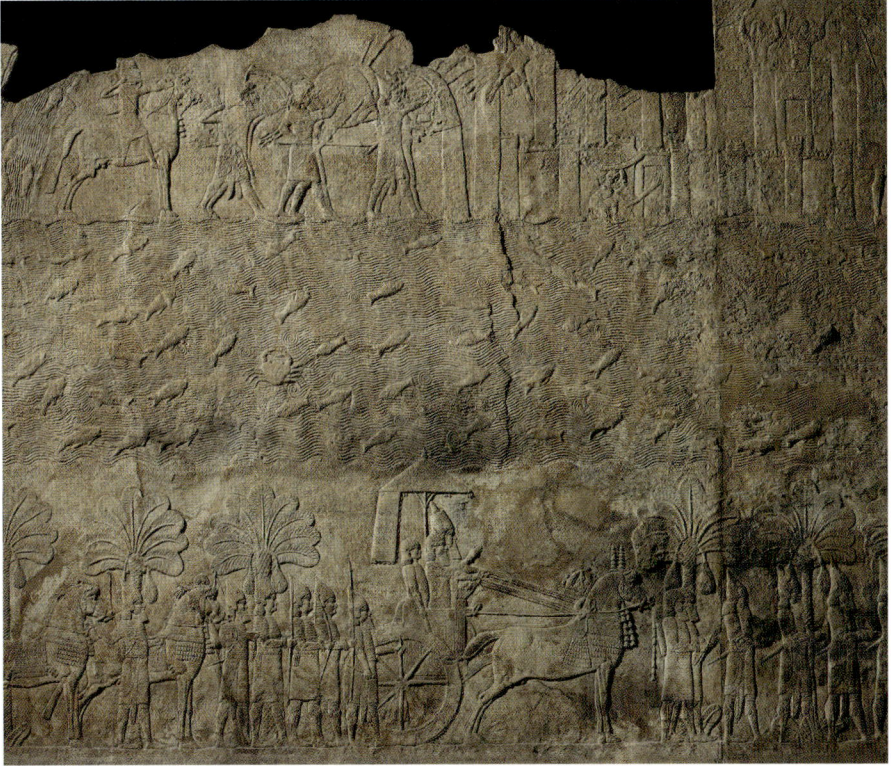

Sculptured relief panel from Ashurbanipal's Southwest Palace at Nineveh showing a row of Assyrian soldiers in front of a palm grove, c. 640–620 BCE
© The British Museum, BM 124825,a

"If the patient has stomach troubles, his nose is bleeding constantly, he has flatulence and suffers from constipation, grind date peel, mix it with suet, make a finger-size suppository, introduce it into the anus, then the patient will get well." [10]

Concerning the use of date pits Dioscorides (I, 109) writes that the burnt pits are astringent and heal wounds; as for the date palm frond, he mentions that a decoction is suitable for kidney and bladder afflictions, for diarrhoea and for unusual vaginal discharge. The dried and ground dates served to treat stomach ailments and bladder disorders. Ibn al-Bayṭār (no. 1955) refers to the astringent properties of the spathe sheaths; soft dates are recommended as a stomachic (no. 425). As for the date palm heart, he mentions its use for treating bile, fever, ulcers, sore throat and haemorrhage (no. 512).

FALSE CAROB, SYRIAN MESQUITE

Akkadian *ašāgu* | Sumerian *kiši*$_{16}$

Fruit | Akkadian *harūbu, kabūt summati* "Dove's dung" | Sumerian *ḫarub*

Prosopis farcta

● ● ○

The false carob is strongly connected in Mesopotamian folklore to the underworld. Perhaps its appearance with gnarled branches and sharp thorns led to the idea of extraordinary power. Even more imposing was the root, so long that it was believed to reach the underworld, where the ghosts of the dead resided. Only Enki, the god of magic and wisdom who dwelt in the *apsu*, the sweet groundwater, was able to cut its roots and thorny branches. This led to the belief that the plant could treat afflictions caused by the 'spirit-of-the-dead'. But it also reflects a reality that faced Sumerian farm workers: according to administrative documents from 3rd millennium BCE Mesopotamia, removing the thorny bush from the fields was one of the hardest farm labours. The work could be performed several times during the year, often repeatedly because the root is lodged firmly in the ground and must be removed completely to avoid new sprouting. Carpenters worked with the wood, and Sumerian poets aptly compared the shape of the fruit (a pod) to that of a water bag.

Prosopis farcta is a member of the pea family (Fabaceae) and is widespread from North Africa and the Arabian Peninsula to Turkey, and from the Caucasus to India. It is very common in Iraq, growing in mountain valleys in the lower forest zone, the lower hills, along rivers on the alluvial plains and in wet sandy depressions in deserts. It is a straggling shrub, with a deep, extensive rooting system and thorny branches, forming impenetrable thickets in some places especially along river banks. Its pods are short, to 5cm long, and swollen. In the traditional medicine of the Near East, the pods and roots are recorded as astringent and given for dysentery; the roots have many other medical uses. Today, the wood is harvested for charcoal-making. Seed remains are found at over 50 archaeological sites in drier parts of the Near East, perhaps reflecting the gathering of the plant for fuel, leading to the frequent inadvertent burning of pods still attached to the plant. Seeds were found at the Iraqi sites of Abu Salabikh (3000–2340 BCE) and Nimrud (715–600 BCE).

The individual parts of the plant had different names. The plant itself was called in Sumerian *kiši*$_{16}$, possibly a loan from Akkadian *ašāgu*; the Akkadian

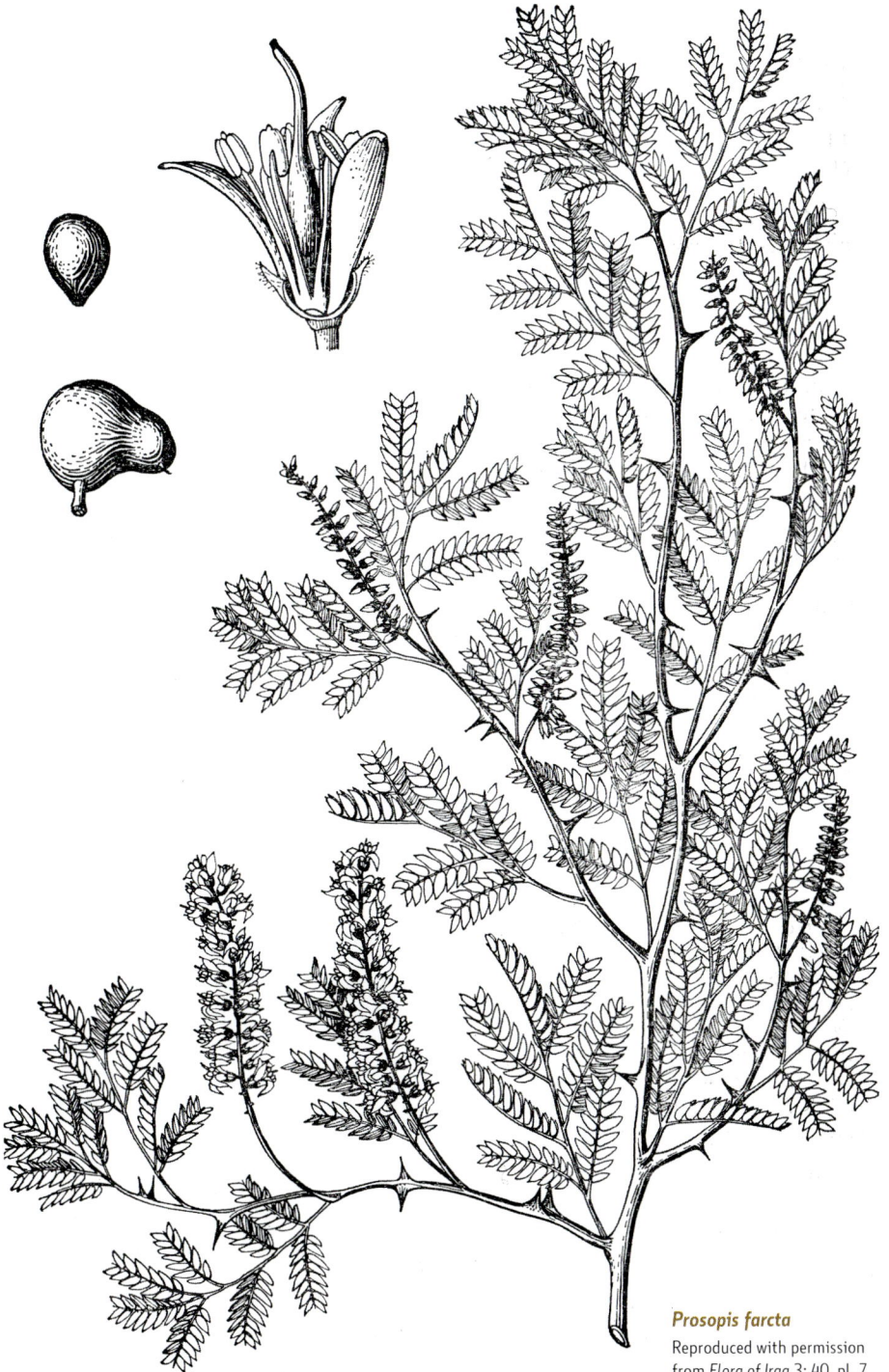

Prosopis farcta
Reproduced with permission
from *Flora of Iraq* 3: 40, pl. 7

term is related to the Arabic name of the plant, *šōk*. The pod was known as *ḫarub* (Sumerian), again a loan from the Akkadian name *ḫarūbu*. The Akkadian term is related to Arabic *ḫurnūb* or *ḫarrūb*. Some Akkadian medical texts use in addition an alias for the seeds or beans, namely "dove's dung" (*kabūt summati*). The English name mesquite goes back to the American Indian term *mizquitl*, which refers to *Prosopis* species growing in North America. False carob, in turn, goes back to Arabic *ḫurnūb* or *ḫarrūb* and comes from the similarities to the pods of the carob tree (*Ceratonia siliqua*). The genus name *Prosopis* is derived from Greek *prósōpon* "face, mask"; the species name *farctum* means "stuffed, filled", referring to the fat pod.

False carob is mentioned in the earliest collection of cuneiform medical recipes, written in the Sumerian language and dating to the end of the 3rd millennium BCE. The practitioner wrote only the recipe down and did not describe the symptoms of the disease. The leaves and further ingredients were processed into a poultice and the root was one of the ingredients of a decoction used to cleanse an affected body part.

Assyrian and Babylonian practitioners in the 1st millennium BCE used the shoots, which could also be stored dried, the leaves, the edible pulp of the pods and the seeds. The powder that was produced from the roasted and ground pulp was employed as a thickening agent in poultices and compresses. The external

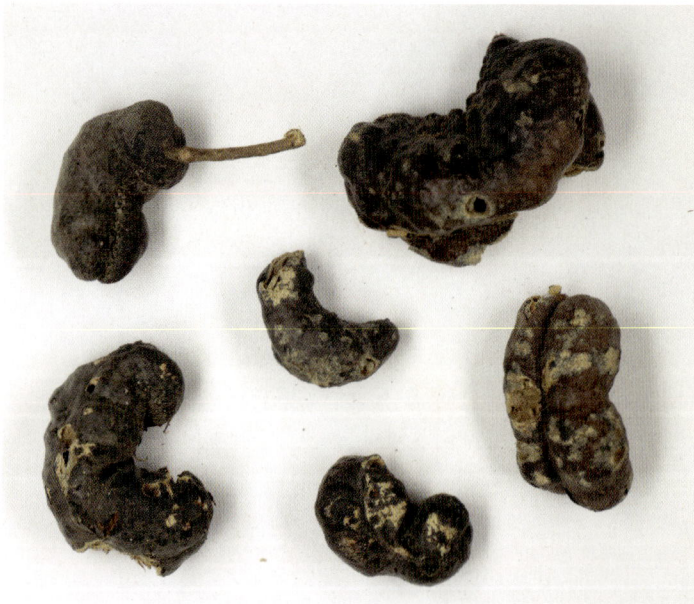

Fruits of Syrian mesquite (*Prosopis farcta*), Bombay, 19th century
Economic Botany Collection, Kew 59154

applications range from the treatment of inflammations of the scalp and hair loss (using the fresh pulp), sores such as gangrene of the feet (using the pulp), to fever (using the leaves and shoots).

"If the patient's head is inflamed and the hair is falling out, in order to remove the inflammation and prevent further hair loss, take seven pods growing on the north side of the tree, roast (the pulp) over burning rushes, mix it with sesame oil, recite an incantation seven times (during the preparation), anoint the patient three times combing him three times. While you are combing the patient recite another incantation three times over his head." [1]

"If a patient suffers from gangrene of the feet and the skin sore is black: crush the fresh false carob pulp, apply it as a bandage." [2]

"False carob leaf – drug for fever heat – to be crushed and rubbed with sesame oil." [3]

"False carob shoot – drug for the demon Lamashtu who causes fevers and chills – to be crushed and rubbed with sesame oil." [4]

The internal applications include urinary tract infections (using the shoots), haemorrhage during pregnancy (using the beans) and toothache (using the root).

"If a man suffers from constriction of the bladder: Crush the false carob shoots, have him drink it with beer." [5]

"If a woman suffers from haemorrhage during pregnancy, in order to stop the blood flow: crush 'dove's dung', she shall drink it with beer on an empty stomach." [6]

"False carob root which the sun god Shamash should not see when you pull it out – drug for the tooth worm (that causes tooth and gum ache) – to be placed on the teeth." [7]

False carob was often used to treat afflictions that were attributed to the 'spirit-of-the-dead'. Spirits had to be revered with regular funerary offerings of food and drink otherwise they would become restless and leave the underworld to haunt the living. According to ancient Mesopotamian belief, spirits entered the body of man through the ears, causing a ringing noise. The symptoms of a ghostly attack ranged from psychological disorders to physical pain and were mainly cured with fumigations and salves. For fumigations, the incense was almost exclusively placed over false carob charcoal. A salve of false carob seed and two more ingredients was prepared in the following way:

"If the 'hand' of the 'spirit-of-the-dead' is lasting long in the patient's body and has not yet been released – in order to remove it: roast two kinds of shells and the seed of false carob over charcoals, crush them into powder, mix them with cedar tar." [8]

Magical cures involved the manufacture of clay effigies representing the attacking spirit, to be buried under the false carob bush. The folk tale was taken up in a magical spell from the beginning of the 2nd millennium BCE:

"O false carob, loyal bush, pure plant, mountain plant –– you do not grow in the rain, you cannot be counted, not even the gods can up-root you. You belong to the god of the underworld Nergal, the one who cares for the 'spirit-of-the-dead'.
O false carob, loyal bush, the lord holds you so tight that only he can cut your root; the great prince Enki holds you so tight that only he can cut your branches. Enki commands over you from the deep fresh-water ocean in the city of Eridu." [9]

FIG

Akkadian *tittu* | Sumerian *peš₃*

Ficus carica

● ● ●

The first cuneiform sources for figs date to the beginning of the 3rd millennium BCE. Throughout Mesopotamian history, fig trees were grown in palm groves, orchards and vineyards. The fruit was traded and sold fresh or dried. The price of the dried figs was comparable to that of raisins. For transport and storage, the dried fruits were tied on strings of 3m length, a practice still prevalent in many countries of southwest Asia.

Fig cake (*Ficus carica*),
Greece, 1903
Economic Botany Collection,
Kew 43209

Ficus carica
Reproduced with permission
from *Flora of Iraq* 4(1): 88, pl. 17

The common fig (*Ficus carica*) belongs to the mulberry family (Moraceae) and is a small shrub or tree, with many branches and large, lobed leaves. The male and female flowers grow on separate plants. The fruit, technically a syconium, contains several hundred minute flowers. In the wild, these are fertilised by a tiny wasp that enters the female syconium, bearing pollen that fertilises the flowers inside which form miniature fruits (fig "seeds"). If the syconium is not fertilised, it will not usually ripen into an edible form. However, parthenocarpic forms of the domesticated fig have been bred that will mature without pollination; these do not contain the characteristic nutty "seeds" of a pollinated fruit. Wild fig trees grow throughout the Near East, the region into which they were taken into cultivation. The domesticated fig is also widespread in cultivation, including in current-day Iraq.

There is a second cultivated fig in the Near East, the sycamore or sycomore fig (*Ficus sycomorus*). Probably first taken into cultivation in Africa, it is a much larger tree, with smaller fruits that are usually gashed with a knife to ensure ripening. Sycamore fig is less cold-tolerant than the common fig, and has been grown in Egypt, Cyprus and the Levant. It is not recorded from Iraq.

Archaeological finds of common fig date back to about 9300 years BCE in the Jordan Valley, near Jericho. There is some evidence these were cultivated,

Sculptured relief panel showing a landscape with, left to right upper row: vine, pine, fig, pine, pomegranate, and lower row: unidentified tree, pomegranate, pine, vine; from the Southwest Palace at Nineveh, c. 700–692 BCE
© The British Museum, BM 124821

but more secure evidence is in the form of fig seeds found at archaeological sites outside the zone where wild figs grow, suggesting that fig trees were widely under cultivation by 4000 years BCE. Fig seeds occur at over 200 archaeological sites in the Near East, showing the importance of this fruit tree; a few seeds were found at the burnt palace of Nimrud (c. 600 BCE) in northern Iraq. Several parts of the tree are used in traditional medicine in the region. When cut, fig trees produce an abundant acrid latex, which is used externally to treat skin conditions such as warts and wounds. The leaves can also be applied externally. The fruit has many uses when consumed, including for coughs and colds, and as a laxative.

The Sumerian name for the fruit and the tree is *peš₃*, the Akkadian term is *tittu*. The Akkadian name is related to Hebrew *te'enah* and Arabic *tīn*. The genus name *Ficus* given by Linnaeus probably comes from the common Latin *fica*, which may derive from Hebrew *pag* or *paggah*, a name given to the green, unripened figs. The species name *carica* is a geographical term, referring to Caria, a region of western Anatolia where figs were cultivated.

Figs were generally appreciated as a sweet food in ancient Mesopotamian cuisine: a popular dessert in the north of Mesopotamia was a fig cake called *libittum*, literally "brick, block", possibly alluding to the cake's shape and texture. Assyrian and Babylonian healers would principally make use of the leaves of the tree.

Fever and inflammation of the scalp, flickering and bloodshot eyes were treated with a poultice made of the leaves.
"If a patient's head is feverish hot, the eyes flicker and are filled with blood: knead 1/3 l fig leaf with milk, shave the head, apply the poultice – it should not be taken off during three days." [1]

The leaves appear in several compound medicines mainly for treating skin lesions of the scalp and the eyelids.

Fruit and leaves were common ingredients in ancient Greek and Medieval Arabic medicine. Dioscorides (I, 128) recommends the plant for the treatment of throat, bladder and kidney afflictions, chronic coughs, arthritis, toothache, inflammation of the uterus, head sores, skin eruptions and leprosy. Al-Kindī (no. 56) uses the seeds for ulcers and the spleen.

FLAX, LINSEED
Akkadian *kitû* | Sumerian *gu* "Thread"
Linum usitatissimum

● ● ●

Flax (*Linum usitatissimum*) was a textile crop in ancient Mesopotamia, with a number of medicinal uses. Flax plants grow up to a metre high, with blue flowers towards the top of the plant, and narrow leaves along the stem. Flax belongs to the Linaceae family and is cultivated for its stem fibres (flax), seeds (linseed) and oil (linseed oil). Today, it is cultivated on a small scale in Iraq for its oil.

Like all crop plants, flax derives from a wild plant. Flax was one of the early crops to be domesticated in the Near East, by 6500 years BCE, from its wild ancestor *Linum bienne*. Seeds are abundant in the archaeological record, being

Linum usitatissimum
Library & Archives, Royal
Botanic Gardens, Kew

found at over 75 archaeological sites in the Near East. This includes finds at several sites in Mesopotamia, including Tell es-Sawwan (6000–5500 BCE), Choga Mami (5300–5000 BCE), Arpachiyah (c. 6000–5000 BCE) and Nimrud (c. 600 BCE). Flax seeds are widely used in traditional medicine in the Near East, most often as a laxative, as a treatment for respiratory conditions such as coughs, and externally as an oil or poultice.

Like many pulses and cereals, flax was a winter crop. It is recorded as a crop in cuneiform texts from the end of the 3rd millennium BCE. There is little evidence that the plant was cultivated as a source of vegetable oil – Mesopotamians preferred sesame oil to that of linseed, possibly because linseed oil cannot be stored for long and easily goes rancid given the storage and climatic conditions of ancient Mesopotamia. Flax was grown instead for its fibres for the manufacture of linen. However, the plant was of minor importance in the textile industry, with wool accounting for about 90% of total textile production. Flax is a nutrient-demanding plant, so that the fields in which it grew had to lie fallow for several years afterwards to recover. It was harvested by hand by a special workforce and required much effort to extract the fibre from the stems. After drying the plants, the seed-bearing capsule had to be removed with a comb. Then the fibres were immersed for some time in water (retted), and beaten to remove the remaining woody parts, after which they were sent to weaving workshops. These factors explain why linen was considered a luxury item and worn only by kings, high-ranking individuals and priests. Linen cloth clad divine statues and decorated tables and thrones – and was also used for medical bandages.

It is not clear whether the Akkadian term *kitû* is a loanword from the Sumerian *gada* or vice versa. While the Sumerian *gada* refers to linen as a textile, the Akkadian *kitû* means both plant and fabric. The Sumerian word for flax is *gu*, literally "thread", alluding to the fibres of the plant. Similarly, the English term flax originates in the Proto Indo-European root *plek-* "to plait" and refers to the manufacturing process of the fibres. Linseed, in turn, is derived from "seed" and Latin *linum* or Greek *línon*, which are of unknown linguistic origin. The Arabic name *kattān* is related to the Akkadian word.

Assyrian and Babylonian practitioners used different sizes of linen cloth as pads to rub in medicines, dressings to protect wounds, wads of absorbent materials for ears, nose, vagina and rectum or simply for bandages. Linseed is mostly used in compound medicines taken as a potion or applied as a poultice and in wadding. During a multiphase treatment of fever, practitioners recommended fumigation by flax seed over charcoal.

Seed of linseed (*Linum usitatissimum*), India, 1934
Economic Botany Collection, Kew 64645

In two texts the plant was used as a single ingredient. Both cases refer to the preparation of a poultice based on milk. The first recipe is for a condition that was simply called "sick buttocks"; under this term, practitioners included a variety of diseases that are linked to the blockage of urine or stool, and severe pains that could radiate to other body regions.

"If a patient suffers from 'sick buttocks', grind 5l of linseed, sift it, make a poultice with milk, apply it to chest and shoulder during 14 days, then the patient shall get well. These are proven ingredients." [1]

The second recipe is for a condition called "stroke" (Akkadian *mišittu*), which seems to have a similar notion to English *stroke*. Often the practitioners mentioned the body part that was affected most.

"If a patient suffers from the consequences of a 'stroke' of the back, grind ½l linseed, heat it in milk in a clay pot, smear it still warm on leather, apply it from the upper arm to the fingers." [2]

Dioscorides (II, 103) recommends linseed as a single ingredient for clysters (enemas) to bring relief from intestinal and uterine pains and as a hip bath for uterine inflammation. The use in compound medicines ranges from the treatment of internal and external inflammation, skin eruptions and chest diseases. Al-Kindī (prescription 110) uses linseed together with raisin, pine seeds and liquorice to treat cough caused by catarrhs. Ibn al-Bayṭār (no. 933) includes references to the use of flax seed for chest conditions, for ulcers of the intestines, and to calm pain.

GARDEN ROCKET

Akkadian *egengiru, gingiru, girgirû* | Sumerian *nig$_2$-gan$_2$-gan$_2$*

Eruca vesicaria (formerly *Eruca sativa*)

● ○ ○

The peppery mustard taste of the leaves of garden rocket led to its culinary use in Mesopotamia from at least the beginning of the 2nd millennium BCE, enhancing meat and vegetable dishes with its pungent flavour. Both the leaves and to a lesser extent the seeds had medicinal uses.

Eruca sativa
Library & Archives, Royal
Botanic Gardens, Kew

Garden rocket (*Eruca vesicaria*) is an annual plant, native to North Africa, the Mediterranean region, and from the Arabian Peninsula to Turkey and China. It belongs to the mustard family (Brassicaceae). The leaves have irregular margins, and the 4-petalled flowers are yellow or white with prominent brownish-purple veins. Garden rocket is found in the lower hills and alluvial plains of Iraq and is also cultivated for its pungent edible leaves.

Archaeological finds are rare, doubtless reflecting consumption as leaves, which generally do not survive in the archaeological record. The seeds have been found at Iron Age sites dating to the 1st millennium BCE in the Caucasus. Leaves of garden rocket have many traditional uses in the Near East today, most often as an aphrodisiac.

The Akkadian name for the plant is spelled variously *egengiru*, *gingiru* or *girgirû* and is related to Arabic *ǧirǧīr*, Aramaic *gargīrā* and Hebrew *gargīr*. The Sumerian name is a compound of *nig₂*, literally "thing", and *gan₂*, literally "field". It is not clear whether Sumerian *nig₂-gan₂-gan₂* is an adaptation of the Akkadian word or vice versa. The English word rocket comes from Latin *eruca*, which has the meanings "colewort" and "caterpillar", related to the Proto-Indo-European root **ghers-* "to bristle", alluding to the shape of the leaves of the plant.

Ancient Mesopotamian practitioners knew the plant especially as an aphrodisiac, as well as a laxative and emetic, and termed it "plant for impotence" and "plant for purging through mouth and anus". They also treated anal sores, skin afflictions and eye diseases with garden rocket. Although it was one of the

Seeds of garden rocket (*Eruca vesicaria*), Saharunpore, India, 1886
Economic Botany Collection, Kew 67333

Garden rocket. Dioscorides Pedanius of Anazarbus, *De materia medica* fol. 49r, Constantinople, mid-10ᵗʰ century
© The Morgan Library & Museum, MS M.652, purchased by J.P. Morgan

principal plants to treat impotence, in practice healing experts seemed to have preferred using it in compound medicines rather than as a single ingredient. The following prescription is one of the few that gives specific details for impotence. It is embedded in a magic ritual that involves an invocation to the goddess of love, Ishtar.

"If a man is impotent when approaching a woman, be it because of old age, illness, fever or accident – in order to make him potent again so that he can make love to a woman: black bryony, plantain, garden rocket and four other (unidentified) ingredients. Crush these seven ingredients, sift them, place an incense burner with juniper before Ishtar, offer a beer libation to her and recite the incantation 'Why did you keep me from making love as if I would be on a road that is suddenly closed off' seven times. Have him drink the ingredients with wine during three days and on the fourth day he shall be fine." [1]

Mixed with lard, the plant was applied to the sores caused by a disease that affected the anus and the buttocks.

"Garden rocket – plant for an inflamed anus – to be mixed with lard and applied to the anus." [2]

"*Garden rocket – plant for a sick anus whatever is the cause – to be mixed with lard and applied to the anus.*" [3]

Most of the prescriptions do not refer to the plant part used, but the following one is an exception. It prescribes the use of seeds in a compound medicine for dark lesions on the face. The use of cedarwood oil possibly promoted the healing of the skin.

"*If a patient's face is covered with dark lesions, crush the seed of garden rocket together with three further ingredients, mix them with cedarwood oil, cool the patient's face with it, then he shall recover.*" [4]

Mixed with mountain honey and a lubricant, it was used for treating the eyes.

"*If the patient's eyes are bloodshot, mix garden rocket with mountain honey and a mineral lubricant, grind the ingredients into a paste and daub the patient's eyes with it.*" [5]

Garden rocket was a well-known aphrodisiac in antiquity; both Galen and Dioscorides (II, 140) report that it generated semen and stimulated the sexual drive. However, the plant also causes headaches, which led the two physicians to caution against over-consumption. Dioscorides (II, 140) mentions that the seeds are used as a diuretic, digestive and laxative. According to Pliny the Elder (XX, 49), the plant improved eyesight. Al-Kindī (no. 60) refers to the uses of the seeds in a remedy for insanity and as a stomachic. Ibn al-Bayṭār (no. 473) refers to the uses of garden rocket for sexual arousal, to increase the appetite and to treat facial sores.

GARLIC

Akkadian *šūmu* | Sumerian *šum*

Allium sativum

● ● ●

Mesopotamian farmers cultivated garlic (and onion) from at least the middle of the 3rd millennium BCE, intercropped among date palms and, as in Iraq today, profiting from the shade of the trees. During the 3rd and 2nd millennia BCE, garlic was considered a luxury food in the Mesopotamian diet, kept for the royal table and that of high dignitaries.

Allium sativum

Garlic (*Allium sativum*) belongs to a large group of bulb-bearing plants in the onion family (Amaryllidaceae), such as onions, leeks and chives, that are native to western and Central Asia. It is an annual bulbous plant, similar to the onion, but the garlic bulb is composed of a number of fleshy scales (cloves), which can be separated and planted to produce entirely new plants. Cultivated garlic flowers are sterile and produce no seed; garlic is therefore always propagated vegetatively. The flowering stem (scape) is sometimes coiled in the upper part, and is surrounded by leaf sheaths for more than half its length. Numerous pinkish flowers are produced in round heads at the top of the stem. Garlic is cultivated throughout Iraq.

The ancestry of garlic is not fully known, but garlic and onion were probably first domesticated in the mountainous regions extending between Turkey and Central Asia. It has been cultivated for at least 6,000 years. Archaeological finds of domesticated garlic depend on unusual circumstances of preservation, such as tomb finds or a burnt kitchen. Desiccated cloves have been recovered from

Bulbs of garlic (*Allium sativum*), Berar, India Museum (before 1879)
Economic Botany Collection, Kew 36699

Egyptian tombs dating back to 3200 BCE, and 150 carbonised garlic cloves were uncovered from the site of Tell ed-Der in eastern Iraq, dating around 1630 BCE.

Garlic bulbs have a wide range of medicinal uses in the Near East, including uses as an external and internal treatment for bites and stings, and they are consumed to treat high blood pressure and rheumatism, and as a decongestant.

The Akkadian name for garlic, *šūmu*, comes from Sumerian *šum* and has entered the vocabulary of Arabic and Hebrew as *ṯum* and *šūm*, respectively. The meaning of the original Sumerian word is unknown. The genus containing garlic was given the Latin name *Allium* (probably from Celtic *all*, meaning pungent or burning). The name garlic is derived from "garleac" – "gár" meaning spear from the spear-shaped cloves, and "léac", an Old English root word meaning plant or herb.

Babylonian and Assyrian practitioners appreciated the plant as a versatile remedy, which explains why it was planted in the first known physic garden in history, that of the Babylonian king Merodach-Baladan II (722–710 BCE). It was used for external and internal applications. Mixed with syrup and sesame oil it was rubbed on the body to lower the body temperature during fever.
"Garlic – drug for fever caused by heat stroke – to rub with syrup and sesame oil." [1]

Made into a poultice it brought relief from pain in the hips. Babylonian and Assyrian healers did not specify what caused the pain; their prescriptions emphasise the preparation of the medicine.
"If a patient's hips are sick: crush garlic, mix it with the roasted flour, knead it together with brewer's yeast, tie it around the hips, the patient shall get well." [2]

The spectrum of internal applications ranges from the treatment of hearing loss, use as diuretic and for flatulence, to the killing of intestinal worms. It was also employed to provoke vomiting. The garlic is usually crushed, but two instances require the consumption of the entire garlic clove, and the use of the squeezed garlic. For a patient who was hard-of-hearing either in the right or left ear, the following was recommended:

"Squeeze garlic, apply it inside the ear, then the patient shall recover." [3]

In case of flatulence and digestive problems, Assyrian and Babylonian practitioners prescribed nearly the entire garlic head.

"If a man's bowels are repeatedly bloated, he shall swallow seven and seven garlic (cloves)." [4]

Possibly, because of its diuretic properties, garlic was used for troubles related to the stricture of the urethra.

"Garlic – drug for narrowing of the urethra – to crush, to drink in sesame oil and first-quality beer." [5]

If the affliction was located in the liver–bile system, ancient practitioners resorted to more drastic measures.

"If a man's bile is caught: crush garlic, he shall drink it with water and vomit." [6]

To clean the patient from intestinal parasites, garlic was taken with beer.

"If a man is sick with intestinal parasite: crush garlic, he shall drink it with vinegar, then it will be expulsed." [7]

The cuneiform handbook of medical recipes included a chapter on the treatment of the physical consequences of being bewitched, which were often related to indigestion and impotence.

"If a man is bewitched: crush garlic, he shall eat it with flavoured sesame oil or drink first-quality beer, suck on lard, then he shall recover." [8]

Dioscorides (II, 152) recommended garlic in case of flatulence, bowel upset, stomach troubles and to expel intestinal worms. He prescribed it for its diuretic properties and to set the urine in motion. Externally, he referred to the use of plasters and poultices for skin afflictions, pustules, leprosy and animal bites. Al-Kindī (no. 59) employs garlic for aches caused by inflammations, pus or fistulas. Among the references that Ibn al-Bayṭār (no. 453) gives for the uses are toothache, colic, scorpion bites and back and hip pain.

GREATER PLANTAIN, RIBWORT PLANTAIN

Akkadian *lišān kalbi* "Dog's tongue" | Sumerian *nig₂-gidru sipa* "Shepherd's staff"

Plantago major or *P. lanceolata*

● ○ ○

It is hard to say whether the Akkadian and Sumerian name referred to broadleaf or greater plantain (*Plantago major*) or ribwort plantain (*P. lanceolata*), or to both. In any case, the roots of dog's tongue were much used in Mesopotamian medicine.

The greater plantain is a perennial plant in the Plantaginaceae family. It lacks a stem and has a short rootstock with many fibrous roots. The long-stalked ovate leaves are 20–40cm in length. The greenish flowers are in a cylindrical head on a long stalk (up to 40cm) that rises above the leaves. The fruit is a small capsule containing up to 30 angular seeds. Its native habitat is in temperate regions of Asia, Europe and North Africa, but it has been introduced and naturalised throughout the world except in dry deserts. Ribwort is similar in appearance but has lance-shaped ribbed leaves. Both plantain species are common plants found throughout Iraq except in extreme desert areas. They grow along roadsides, footpaths, waste places, in shady orchards, banks of shallow pools and streams and in wet margins of water seepages.

The seeds of *Plantago* are distinctive and easy to recognise in archaeological samples, but usually cannot be identified to species. They have been found in the archaeological plant remains from over 50 excavations in the ancient Near East, doubtless reflecting their prevalence as a weed of disturbed ground and

Seeds of greater plantain (*Plantago major*), Lahore, India Museum (before 1879)
Economic Botany Collection, Kew 45995

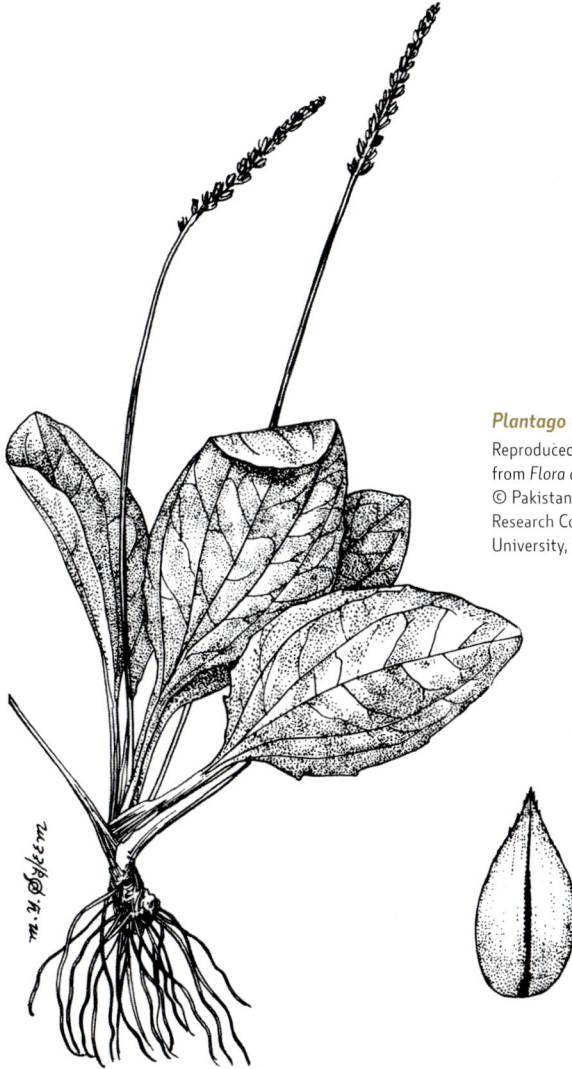

Plantago major
Reproduced with permission
from *Flora of Pakistan* 62: f. 2
© Pakistan Agricultural
Research Council and Karachi
University, Pakistan.

field edges. A single seed was found in the Neo-Assyrian levels (715–600 BCE) of
Nimrud, northern Iraq.

Both plantain species are widely used in the Near East today. The leaves have
an astringent effect and are often applied externally to skin problems such as
warts, infections and stings. In Turkey, Greece and neighbouring countries,
a decoction or tea made from leaves is taken internally for many illnesses,
including dysentery, diarrhoea, chronic constipation and chest diseases. The
seeds are used less often, in Iraq to make poultices for boils and as a treatment
for bowel problems.

The appearance led to the various ancient names for plantains. For Babylonians and Assyrians, the leaves were the basis of names. They called plantain "dog's tongue" (Akkadian *lišān kalbi*), while ancient Greeks called it in turn "lamb's tongue" (*arnóglōsson*), a term adopted in the Syriac and Arabic languages (*lešān ʾemra* and *lišān al-ḥamal*). The Sumerians, on the other hand, considered the erect growth of the plantain flowers so characteristic that they compared it with the stick shepherds used to manage their sheep, referring to it as "shepherd's staff" (nig$_2$-*gidru sipa*). The Latin name *Plantago*, from which the English plantain derives, goes back to Latin *planta*, "foot sole", and refers to the shape of the flat leaves. Since its seeds stick to feet and paws, the plant was spread wherever man or beast would go.

In contrast to current traditional use of the leaves and seeds, Babylonian and Assyrian practitioners mainly used the root of plantain, which is white, soft and about the size and thickness of a finger. A potion of the fresh root helped to alleviate bouts of violent and wet cough. This cough was simply called "phlegm" (*suālu* in Akkadian) and some potions were meant "to strip it off".

"If a patient suffers from 'phlegm': Boil fresh dog's tongue as if it were a turnip, mix it with sweetened milk and pressed oil, have the patient drink it on an empty stomach, then he will recover." [1]

If a patient would suffer from "wind in the belly", as ancient Mesopotamians termed the gases that cause severe bowel troubles and intense colicky pain, the root was to be dried and soaked before being administered.

"If 'wind' keeps moving about in the patient's belly – to heal him: dry the root of dog's tongue which was not exposed to the sun when it was pulled out, crush it, have the patient drink it on an empty stomach with best-quality beer, then he will recover." [2]

The same potion could be used to ease pains if a man suffered from bladder stones and to treat jaundice.

"If a patient's eyes are full with jaundice: dry root of dog's tongue, crush it, soak it in beer, have the patient drink it, then he will recover." [3]

"If a man suffers from calculi, dry dog's tongue which was not exposed to the sun when it was pulled out, crush it, add it to brewer's beer, let it macerate overnight under the stars, have him drink it on an empty stomach, then he will recover." [4]

Another medicine to be drunk, this time for spitting blood, was prepared by squeezing the whole plant.

"Dog's tongue – drug for blood-spitting cough – squeeze out its juice, have the patient drink it." [5]

Plantain. *Kitāb al-Ḥ ashā'ish fī Hayūlā al-'Ilāğ al-Ṭibbī*. **Arabic translation of Dioscorides'** *De materia medica*; **revised edition of Abū 'Abdallāh al-Nātilī (380/999) of the original Arabic translation by Ḥunayn ben Isḥāq (d. 260/873), Leiden, Universiteitsbibliotheek, MS Or. 289, fol. 85r**

The seeds of plantain could be processed into an eye salve to soothe persistent watering eyes.

"If the top of a patient's head is hot, the temples throb and the eyes are so severely affected that they are blurred with a film, a cloud, haziness, a 'rose' or a 'worm' and are persistently weeping: grind the fresh seed of dog's tongue, extract its juice in a bowl, allow to dry. As soon as it is dry, crush it with copper patina in oil, apply it to the eyes." [6]

The leaves of plantain, on the other hand, were used for treating potency problems by crushing them, mixing them with oil and applying them to the penis. *"Plant on whose leaves lizards rest: 'Shepherd's staff' or 'dog's tongue' is its name. It is good for potency problems; crush the leaves, rub him with sesame oil."* [7]

Another way of processing the leaves and root was to make a healing poultice that was effective at relieving animal bites.

"Medicine for snake (bites) – dog's tongue." [8]

Many of the medical applications described in the cuneiform recipes are likewise found in Dioscorides (II, 126). He recommended the different parts of plantain for blood-spitting, eye problems, bladder ulcer, colic and diarrhoea. Virility troubles and jaundice, however, are absent from the list. Medieval Arabic literature, too, included the same areas in their medical practice, as well as impotence and jaundice (Ibn al-Bayṭār no. 2022). Al-Kindī (no. 236) mentions the use of plantain for boils, ulcers, external haemorrhoids and as dentifrice.

HENBANE

Akkadian *šakirû* "(Plant) which intoxicates" | Sumerian *šakir*

Hyoscyamus niger or *H. albus*

● ○ ○

Assyrian and Babylonian healers used henbane leaves to treat the puncture wounds of scorpion stings and snakebites, injuries from dog bites, and open wounds and inflammations caused by gangrene. It also served to alleviate rheumatic pains and was applied as an analgesic to the teeth in case of toothache.

Hyoscyamus niger and *H. albus* are annual or biennial species in the tomato family (Solanaceae), which contains other poisonous and psychoactive plants. Like thorn apple (*Datura stramonium*) and mandrake (*Mandragora* spp.), they contain highly poisonous tropane alkaloids, including hyoscyamine, atropine and hyoscine (scopolamine). Both species of *Hyoscyamus* grow wild in modern-day Iraq in disturbed habitats and cultivated land in the lower hills of northern Iraq. *Hyoscyamus niger* is distributed from Europe eastwards to Siberia, and *H. albus* from Europe eastwards and south to North Africa. *Hyoscyamus niger* has an unpleasant smell and is covered with sticky glandular hairs. *Hyoscyamus albus*, on the other hand, is a pleasant-smelling plant and is covered with long white hairs. The creamy-yellow flowers with purple venation are arranged in two rows at the top of stems – the purple venation being absent in *H. albus*. The fruit in both species is a small capsule enclosed by the sepals when the fruit is mature.

Hyoscyamus niger
Reproduced with permission from *Flora of Pakistan* 168: fig. 11
© Pakistan Agricultural Research Council and Karachi University, Pakistan.

Henbane grows in fields and habitats disturbed by humans, so it is not surprising that its seeds are found in about 25 archaeobotanical assemblages in the Near East, most likely as a component of weed flora. Most reports do not attempt to identify seeds to species. Both species have a wide range of traditional uses in the Near East, with an emphasis on the use of decoctions of the leaves to treat eye problems.

The plant is called in Akkadian *šakirû* and in Sumerian *šakir*, derived from the Akkadian words for "a strong drink, fermented beverage" (*šikaru*) and "to become inebriated, drunk" (*šakāru*). The poisonous and hallucinogenic effect is reflected in the English name henbane, explained as a plant that poisons chickens. The genus name comes from the Greek *huoskúamos* meaning "hog's bean". In Arabic, the plant *H. niger* is called *banj*, which is also used to refer to anaesthetics in general, while *H. albus* is *sakrān*, related to the Akkadian word.

One of the oldest references for the use of henbane in ancient Mesopotamia dates back to the middle of the 2nd millennium BCE. Sometime in the 15th or 14th century BCE, the physician Belu-muballiṭ from the city of Ur wrote a letter to his overlord in which he asked him to send henbane to prepare a potion for severe abdominal colic. The case was grave, not only because Belu-muballiṭ was in charge of the local princess but because the physician feared that the colic could worsen and develop into a far more serious illness called "hand of the curse". Medical practitioners of the 1st millennium BCE knew henbane as one of the principal plants to treat scorpion stings and serpent bites. For treating dog bites, the fresh plant was to be crushed and applied to the sore.

"Fresh henbane – drug for dog bites – to be crushed and applied to the wound." [1]

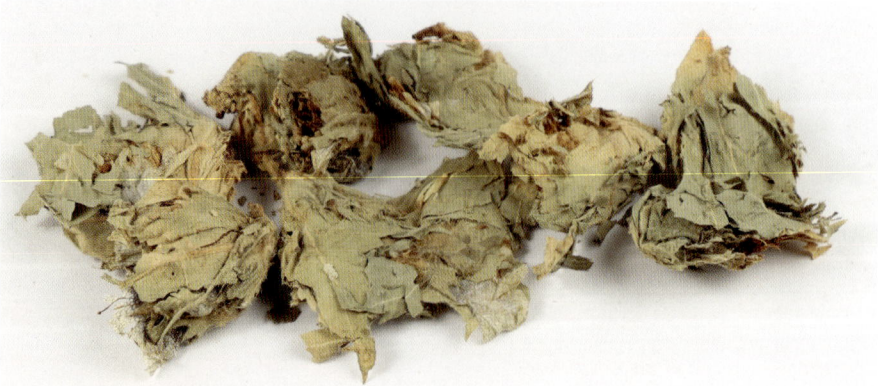

Leaves of henbane (*Hyoscyamus niger*)
Economic Botany Collection, Kew 46879

Other external treatments included the preparation of a salve for a disease that affects muscles and joints, and of a poultice for curing inflammation and sores on the feet.

"The plant is good for muscle and joint disease maškadu. *Dry it, crush it, rub the patient with sesame oil and he shall get well."* [2]

"If a patient suffers from gangrene of the feet, dry henbane, crush it, mix it with coarsely ground malt, prepare it as if making a poultice and bind it on the feet." [3]

The root was much used to treat toothache as is mentioned in the following statement:

"Henbane root – drug for toothache – to be crushed and applied to the tooth." [4]

The root also served to alleviate the pains during a difficult birth.

"Henbane root – drug for difficult birth – to be crushed and applied to the neck." [5]

The seeds of the plant were likewise taken as medicines. The disease to be treated was called "hand of the god", which had many symptoms including depression. It was believed that the personal tutelary (guardian) spirit would send afflictions when a person became estranged from the deity.

"Seed of henbane – drug for 'hand of the god' – to be crushed and washed with water." [6]

Apparently the treatment of abdominal colic in humans was so effective that it was also used in veterinary medicine. Medical doctors seem to have served as veterinary surgeons, with the few existing veterinary prescriptions incorporated into the handbook of human medicine, and thus falling into the remit of the physician. The animal treated was the horse – the care for horses was vital in times of peace and war. It performed essential cavalry roles and, above all, moved the chariot of the king and that of the gods during religious processions.

"It is good for colic in horses. Dry it, crush it, squeeze grapes, pour it into its left nostril and it will get well." [7]

Dioscorides (IV, 68) differentiated between three kinds of henbane, of which one was recommended. The juice of the seeds was used for analgesic pessaries, earache, rheumatism, problems related to the uterus and inflammation of the eyes. He recommended the seeds mixed with coarse meal for inflammations of the feet and the root as mouthwash for toothaches. Al-Kindī (no. 45) uses henbane in a compound medicine to treat cold ailments, epilepsy, insanity and black bile. Ibn al-Bayṭār (no. 356) includes in his chapter on henbane the treatment of cough, teeth problems and colic, as well as its use as a general analgesic that can be replaced with opium.

JUNIPER

Akkadian *burāšu* | Sumerian *šim-li, še-du₁₀* "Sweet grain"
Juniper berry | Akkadian *kikkirannu, zēr burāši* | Sumerian *šim-li*
Juniperus species

● ● ●

In ancient Mesopotamia, juniper wood was used to manufacture wagon poles, scales for heavy weights, furniture and especially bows. The trunks of the tree, which were said to smell sweet, served as roof beams, and the branches and seed cones were traded commodities from at least the middle of the 3rd millennium BCE onwards. Cooks would flavour meals with juniper berries on special occasions at the king's table. Juniper berries also had important medicinal uses, particularly for the treatment of jaundice.

There are two native species of juniper in Iraq: *Juniperus oxycedrus* and *J. polycarpos*, belonging to the juniper family (Cupressaceae). *Juniperus oxycedrus*, commonly called the prickly juniper (because of its stiff spine-like leaves), brown-berried juniper or cade oil plant, is very common in the forest zone of northwest Iraq. The wood of this species is distilled to give the "oil of cade" noted for its medicinal properties, particularly for skin diseases. *Juniperus polycarpos*, known as the Himalayan cedar, is distributed from Turkey to Iran, eastern Arabia, Turkmenistan and northwest Himalaya, but is extremely rare in Iraq, occurring only at the border areas with Turkey. Although the wood and berries (female cones) of the Himalayan cedar are of economic importance, the tree is so rare in Iraq that it is of no economic significance. Two other species of juniper, the Syrian juniper (*J. drupacea*) and the Grecian juniper (*J. excelsa*), both yield valuable wood, and berries that are much used and relished

Fruits of juniper (*Juniperus oxycedrus*), Cyprus, 1882
Economic Botany Collection, Kew
28782

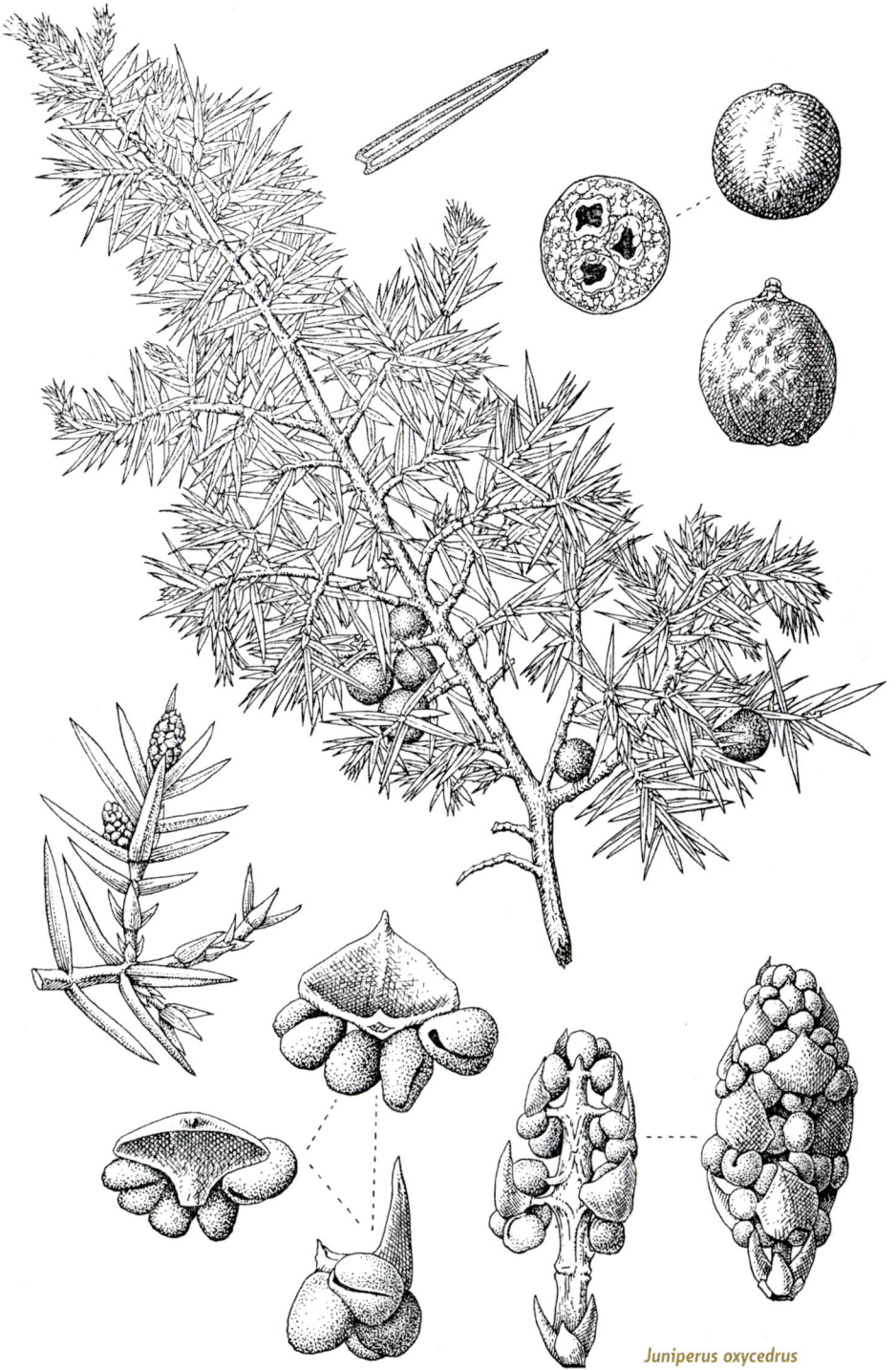

Juniperus oxycedrus
Reproduced with permission
from *Flora of Iraq* 2: 92, pl. 12

in foods and for the oil they produce. The former is native to southern Turkey with scatterings of populations in Syria and Lebanon. It produces walnut-size berries (called in Arabic 'abhal) that are much prized as food. Berries of the Syrian juniper were listed by James Felix Jones as one of the drugs sold in the markets in Baghdad in the mid-nineteenth century. The cuneiform information is insufficient to identify a particular juniper species, and it may be that berries from several species were available by trade in ancient Mesopotamia.

Juniper berries rarely occur in the archaeological record, reflecting that their role as a prized flavouring and medicine led to them rarely coming into contact with fire. Juniper wood is often found in charcoal remains, reflecting its use as fuel, or the burning in destruction fires of juniper roof beams. Probably the best-known find of both wood and berries is from the tomb of Tutankhamun in Egypt, dated to about 1320 BCE. Juniper berries have a wide range of traditional uses throughout the Near East, including treatment of coughs and colds, as a diuretic, for stomach problems and much else.

The meaning of the Akkadian name burāšu is unknown; the Sumerian name for the tree še-du$_{10}$ means literally "sweet grain" and relates to its berries. The Akkadian name of the juniper berry was kikkirannu, which was also used to refer to cones of other conifers. Occasionally, they referred to the berries as "juniper seed". The Sumerian term šim-li refers to both the cones and the tree; it is a compound name consisting of šim "aromatic", which alludes to the spicy scent of the tree and its parts and li, which is of unknown meaning. The English name comes from Latin iuniperus of unclear etymology; it might be related to Proto-Indo-European *ioi-ni- "reeds, rushes".

Assyrian and Babylonian practitioners used both the whole berries, and the flesh extracted from inside. These were crushed or ground to a powder or flour, which could be drunk with beer, or mixed with water or sesame oil. In the private drug stores, the berries were kept separately from the branches. Religious experts employed juniper (wood and branches) for incense, which was burnt before praying to the gods. It is not known whether juniper tar or cade oil were made in ancient times; this requires heating the wood and branches and capturing a distillate – the tar. It is a thick viscous liquid of brown colour that is used in traditional Middle Eastern medicine for the treatment of eczema.

Juniper was one of the principal drugs for treating jaundice, used as salves and potions.
"Juniper – drug for jaundice – to be rubbed with sweet milk and sesame oil." [1]
"If the patient is sick with jaundice, have him drink juniper with beer." [2]

Juniper. *Kitāb al-Hashā'ish fī Hayūlā al-'Ilāğ al-Tibbī*. Arabic translation of Dioscorides' *De materia medica*; revised edition of Abū 'Abdallāh al-Nātilī (380/999) of the original Arabic translation by Ḥunayn ben Isḥāq (d. 260/873), Leiden, Universiteitsbibliotheek, MS Or. 289, fol. 22r

"If the patient's body is yellow, the face is yellow and he loses weight, the disease is jaundice. Crush juniper, have him drink it with beer or milk." [3]

Potions were also recommended for other biliary diseases. Some recipes mention the emetic effect of the medication.

"If the patient's bile is caught, crush juniper, have him drink it with beer so that he vomits." [4]

The same effect was sought when treating abdominal pains using the berries. Whether all potions for intestinal problems were used as an emetic is not clear.

"If the patient's belly is sick and he has abdominal pains, have him drink juniper berry with beer so that he vomits." [5]

"If the patient has colic, crush juniper berries, have him drink it with beer, then he shall recover." [6]

"If the patient's intestines are inflated, he is constantly coughing up phlegm and has fever, crush juniper, have him drink it with beer." [7]

A potion made of finely ground juniper cones was given to women suffering from postpartum belly bloat.

"If a woman has given birth, her belly has become distended and is inflated, grind juniper, have her drink it with first-quality beer, then she shall recover." [8]

The inside of the juniper berry was recommended for the treatment of chest diseases; because of the bitter taste of the berries, they were used as electuary, that is, mixed with something sweet to make the medicine more palatable.

"To keep away a bronchial disease, mix syrup, ghee, filtered sesame oil and the inside of juniper berry residue, have the patient take it by the spoonful." [9]

Assyrian and Babylonian practitioners believed that a potion of the berries would repel snakes and thus protect from snakebites.

"Juniper berry – drug for a specific snake – crush it, sift it and have the person drink it. The snake will not bite the person." [10]

Juniper was employed for discharge from the ears and for skin lesions. For the ear, practitioners prepared a small wad of absorbent material with ground juniper. Juniper berry oil was used as salve for sore skin and for treating fever and a poultice made from the berries for cracked heels.

"If the patient's ears continually ring and they discharge pus – grind juniper, wrap it into fleece and insert it into his ears." [11]

"If the patient's heel is fissured, crush the dry inside of juniper berry, prepare a poultice by making a paste with mustard water that you have warmed in a copper pot, apply it as a bandage." [12]

"Juniper – drug for a skin disease that affects head and eyes – add it to sesame oil, warm it over coals, anoint the patient's head with the warm salve." [13]

"If the patient presents during an infectious disease fever due to an affliction of the eyes which has caused blurred vision, sprinkle repeatedly juniper oil on his temples until the fever of the eyes has disappeared." [14]

It seems that practitioners were aware of the beneficial properties of juniper for the teeth and gums as they prescribed it for brushing the teeth.

"Inside of juniper berry – drug for brushing one's teeth – rub it on the teeth on an empty stomach." [15]

Dioscorides (I, 75) includes a short paragraph on juniper berry: it is useful for the stomach, bronchial diseases, flatulence, colic and ulcers. It serves as a diuretic and is helpful for uterine troubles, spasms and ruptures. Ibn al-Bayṭār (no. 1528) refers to the use of juniper for chest and liver ailments, epilepsy and poisonous animal stings.

LEEK

Akkadian *karašu* | Sumerian *garaš*
Allium ampeloprasum (formerly *A. porrum*)

● ● ●

Alongside onion and garlic, leek was cultivated in garden plots in ancient Mesopotamia by the middle of 3ʳᵈ millennium BCE, often intercropped among date palms and other fruit trees. Some of these plots covered up to 34ha. The Sumerian and Akkadian literature compares leeks to the hair of deities. The ancient Mesopotamian goddess of the Netherworld, Ereshkigal, for example, is said to have the hair on her head bunched up as if it were leeks and the fibrous root of the plant was compared to armpit hair.

Like onion and garlic, leek is a bulbous herb in the onion family (Amaryllidaceae). The scape (stem) is up to a metre long, covered for about a third of its length with leaf sheaths. The leaves are flat but keeled below. The many pink to white flowers are borne in a rounded head. Wild leek is found in Iraq on the lower hills and as a weed among crops; it differs from the cultivated variety in having a distinct bulb and numerous bulbils, and narrower leaves. There are no archaeobotanical records of leek in the Near East.

Onion, garlic and leek were subject to food taboos, especially on days of the cultic calendar when important religious ceremonies were celebrated, namely at the beginning of the first and the seventh month of the Babylonian year, *Nisannu* (March/April) and *Tašrītu* (September/October). The meaning of the Akkadian term *karašu* is unknown; from Semitic Mesopotamian, it entered Sumerian (*garaš*), Arabic (*kurrāṯ*) and Hebrew (*kāreš*). The English name "leek" is related to the Proto-Indo-European root *h_2el-* "to grow".

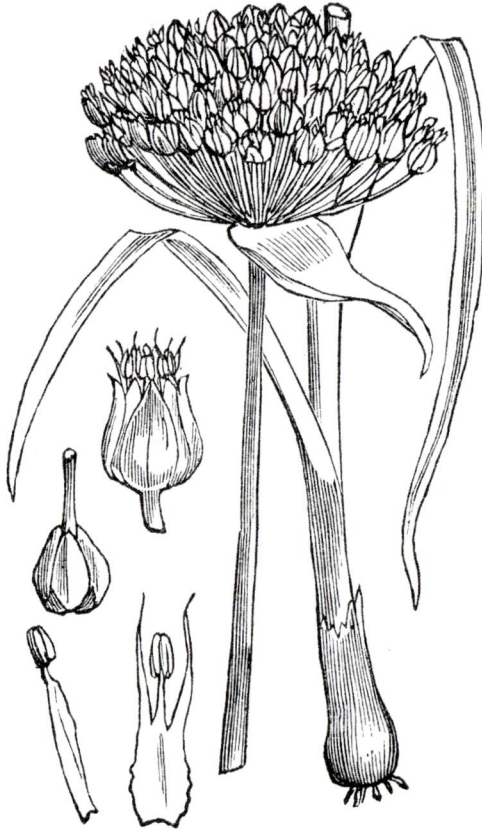

Allium ampeloprasum

Leek was not only appreciated as a vegetable at the ancient Mesopotamian table but also used for its medicinal properties. However, Assyrian and Babylonian practitioners warned that leek would diminish the vision and even cause an eye disease.

"If the patient eats leek, the eyesight will diminish." [1]

"The person must not eat the leek root or else he will get the qūqānu *eye disease."* [2]

They prescribed the plant for intestinal troubles, problems of the womb and a disease called in Akkadian *bu'šānu* that affects the nose, mouth and throat causing bleeding, excess salivation and fever. The plant was usually boiled and then applied externally to the affected area or drunk as a potion. In the case of stomach problems, the practitioners used pickled leek.

"If a patient's belly is sick: mash pickled leek, he shall drink it with milk." [3]

To expel intestinal worms the healing expert had *"the patient eat carefully leek and have him drink hot beer, then the patient will groan and produce the worm."* [4]

Leek. *Kitāb al-Ḥashāʾish fī Hayūlā al-ʿIlāǧ al-Ṭibbī*. Arabic translation of Dioscorides' *De materia medica*; revised edition of Abū ʿAbdallāh al-Nātilī (380/999) of the original Arabic translation by Ḥunayn ben Isḥāq (d. 260/873), Leiden, Universiteitsbibliotheek, MS Or. 289, fol. 92r

Leek was rubbed in the mouth to cleanse the respiratory organs, by bringing relief from excess salivation or phlegm and from the sensation of a hot and inflamed nose.

"If a patient's nose is hot, he suffers from excess salivation, the disease called bu'šānu *has caught him: boil leek, after some time apply it by rubbing to the patient's mouth."* [5]

The following prescription is meant for women who have just given birth. It shows that leek was also used to reduce fever and to treat skin eruptions.

"If a woman has given birth and she is hot, pustules and black spots appear on her body and her muscles, her inside holds pus – in order to calm it down: cool off the milky sap of leek, rub her and introduce it into her vagina." [6]

Dioscorides (II, 149) mentions that leek was used for the stomach and the chest; it stopped nose bleeds, was applied to swellings, pimples and wild animal bites and used for treating problems of the womb. He also warns, like the Assyrian and Babylonian practitioners, that leek diminishes the eyesight. Al-Kindī (no. 255) prescribes leek for headache and to treat haemorrhoids. Ibn al-Bayṭār (no. 1910) refers to the use of leek for colic, digestive troubles and problems of the liver and spleen. The leaves were considered useful to prevent abortion.

LIQUORICE

Akkadian *šūšu* | Sumerian *še-nu*, *še-du₃-a* and *šušu*

Glycyrrhiza glabra

● ● ○

Several species of the genus *Glycyrrhiza* are a source of liquorice (also spelled licorice) but in the Middle East *Glycyrrhiza glabra* is used. This species belongs to the bean family (Fabaceae) and grows wild in dry and semi-dry areas from southern Europe to Pakistan and Central Asia, and beyond to Mongolia and China. In the Middle East, it grows wild and is also cultivated as a medicinal herb. In his diary, T.E. Lawrence (1911) records the cultivation of liquorice in Syria: "olive-yards and vineyards and fields of liquorice". It was an important medicinal plant in ancient Mesopotamia.

Glycyrrhiza glabra
Reproduced with permission from *Flora of Iraq* 3: 447, pl. 87

Root of liquorice (*Glycyrrhiza glabra*), Iran, 1947
Economic Botany Collection, Kew 57838

The plant is a woody perennial, growing up to 60cm tall. The dried roots (rhizomes) are the source of liquorice; they are hard and fibrous, about 1cm in diameter, with brown skin and a yellow interior. The feathery leaves are made up of smaller oval leaflets, 1–4cm long and 1–2cm wide. The small flowers are blue to purple and arranged along stalks about 20cm long. The pod-like fruit is about 2cm long and contains 1–6 seeds. Liquorice is widespread in the lower hills of northern Iraq and in the alluvial valleys of the south. Given that the root is the useful part of the plant, it is not surprising that the seeds of liquorice have only been found at a handful of archaeological sites.

The raw root is used worldwide for flavouring. In powdered form, it is used in sweets, baked goods, ice cream and soft drinks, and medicinally in cough mixtures and lozenges. A tea made from the roots is an excellent thirst-quencher and the powdered root is also used as a sweetener in herbal teas. In Mongolia, the leaves are used as a tea substitute. Liquorice roots contain a triterpene saponin, glycyrrhizin, which, being 50 times sweeter than sugar, is responsible for the sweet taste of the roots. Glycyrrhizin is broken down in the body to a chemical with mildly anti-inflammatory properties. The roots have a very wide range of traditional uses in the Near East, including as a demulcent in the treatment of sore throats and an expectorant in the treatment of coughs and bronchial catarrhs, reducing fevers, and treating inflammations such as chronic skin conditions, arthritis and rheumatic diseases. Kidney disease and gastric ulcers also figure in traditional Middle Eastern medicine today as afflictions that are commonly treated with liquorice root.

The Akkadian *šūšu* and the Sumerian terms *še-nu*, *še-du₃-a* and *šušu* belong to the group of plant names that we cannot currently explain. From Semitic Mesopotamia, the term entered Arabic (*sūs*) and Hebrew (*šūšah*). Ancient

Greeks called the plant *glukúrhiza*, a contraction of *glukús* "sweet" and *rhíza* "root". The English name goes back to Latin *liquor* "liquid" and *succus liquiritae* "liquorice juice", which is a corruption of the ancient Greek name. The species name, *glabra*, means glabrous, that is, smooth, without hairs, referring to the hairless pods.

Liquorice grows abundantly alongside the Euphrates, under date palms, and it was easily available in Assyrian stores of *materia medica*. The oldest cuneiform medical recipes recommending liquorice were written around the beginning of the 2nd millennium BCE. However, the use of liquorice as a medicinal plant is surely older.

Assyrians and Babylonians knew liquorice not only as a medicinal plant, but also as a dessert eaten with fruit. When the great Assyrian king Ashurnasirpal II (883–859 BCE) had completed his palace in the ancient city of Nimrud, he celebrated with a huge banquet, reportedly for 69,574 guests. He offered about 800 litres of liquorice alongside almonds, pistachios, dates and other snacks. A rough calculation suggests each guest could have delighted themselves with two rations of liquorice. Assyrian and Babylonian practitioners occasionally used another term for liquorice, "mongoose tail". It is not clear whether this is a secret name that only they understood or one that describes the appearance of the plant. Indeed, with a little bit of imagination, the freshly dug-up root of liquorice resembles in colour, length and thickness the tail of this small carnivorous animal.

In cuneiform medical recipes, liquorice was taken to treat coughs, jaundice, troubles with the bile, abdominal pains and inflamed intestines. Used externally, it lowered fever, served as a mild laxative, provoked vomiting and was used for skin problems. As a powder it was applied to swellings in the ear, and processed into pills, it promised to increase the sexual response of women.

One of the oldest recipes written down in the Akkadian language (dating to the beginning of the 2nd millennium BCE) describes how jaundice was treated with a potion of liquorice root.

"If a man suffers from jaundice, soak liquorice root in milk, let it macerate overnight under the stars, mix it with cold pressed oil, have him drink it and then he shall recover." [1]

The medication and preparation method remained similar in subsequent periods. Cuneiform recipes from almost 1,000 years later would substitute the milk with beer, and use as an alternative to the extraction process of cold, overnight maceration, that of warm oven maceration, and recommend not only taking the medication in oil but also in beer or a mixture of both.

Less commonly, the juice of the root could be prescribed for jaundice and the bark of the root used for treating problems associated with the yellow–brown bile. *"Drug liquorice root – drug for jaundice – dig out the root before sunrise, mince it, squeeze out its juices, have the patient drink it."* [2]

"Drug bark of liquorice root – drug for bile – to be taken with beer." [3]

Jaundice refers to the yellow colouration of the skin and the "whites" of the eyes, caused by an excess of the bile pigment bilirubin in the blood; bile fluid is produced in the liver and stored in the gallbladder. The excess may result from liver disorders or from obstruction of the bile ducts. Ancient Mesopotamian experts certainly did not know what caused the discolouration of skin and eyes; they associated bile troubles with the discolouration of skin and eyes because of the colour of bile.

Assyrian and Babylonian healers prescribed a potion of liquorice root for bouts of violent coughing, called simply "phlegm" (in Akkadian *suālu*), and for blood-spitting, as well as for abdominal pains and inflamed intestines.

"Drug liquorice root – drug to stop productive cough and for coughing up blood – to be taken with beer." [4]

"If a patient suffers from abdominal pain called 'Binding of the belly', have him drink the root of liquorice with water on an empty stomach." [5]

"If the patient's intestines are inflamed and the feet are swollen, soak liquorice root in water, let it macerate overnight under the stars, filter it, have the patient drink it on an empty stomach and he shall recover." [6]

Several cuneiform recipes describe the external use of liquorice leaves. They were crushed and processed into a salve to treat a skin condition with reddish clustered lesions, febrile headaches and a runny nose. In the form of a poultice (bandage) or simply sprinkled over the affected part, the leaves were applied to gangrene of the feet.

"If a patient suffers from gangrene of the feet and the surface is black, crush fresh liquorice, bandage the foot." [7]

"If it is gangrene of the foot, crush liquorice leaves, distribute them over the surface of the afflicted part." [8]

The same poultice was used for a feverish sensation in the belly, with the intended effect of reducing the fever.

"If a patient's belly is feverish hot, mix liquorice leaves and flour, make a poultice with mustard water, apply it to the belly." [9]

An exceptional external use of the root is the magical ritual that was performed to enhance the libido of a woman. While the incantation expert recited several times the same spell, he manufactured small liquorice pills:

Liquorice. Dioscorides
Pedanius of Anazarbus,
De materia medica.
Constantinople,
c. 512 CE. Cod. Med
Gr. 1 (facs.), fol. 91v
© Österreichische
Nationalbibliothek, Vienna

"Take pulverised liquorice root at sunrise or sunset, form it into seven and seven pills, recite the incantation seven and seven times over them, smooth the pills a second time and recite the incantation again, then place them between her breasts and your wife will come to you." [10]

A faint echo of this practice is found in the modern traditional medicinal practice of Iran and Iraq to arouse sexual desire. In Arabia, the roots are also used with other ingredients to increase sexual potency.

Theophrastus (371–287 BCE; Book IX, 13.2) mentions the application of liquorice root for chest ailments, to expel phlegm or to alleviate a sore throat. Dioscorides (III, 5) states that both a decoction of the fresh root and the juice of liquorice, drunk with grape syrup, would treat liver and kidney ailments, itching of the bladder and heartburn. Al-Kindī (no. 159) and Ibn al-Bayṭār (no. 1250) recommend liquorice for jaundice and diseases of liver and spleen, as a remedy for wounds and ulcers, as well as for diseases of the respiratory tract, stomach, kidneys and bladder.

ONION

Akkadian *šamaškillu, šusikillu* | Sumerian *šum-sikil* "Pure white bulbous plant"

Allium cepa

● ● ●

Evidence of the cultivation of onions appears in early cuneiform administrative documents from the middle of the 3rd millennium BCE onwards. Together with leek and garlic, onions were grown in garden plots, often intercropped among date palms. During the 3rd and 2nd millennia, onions were a luxury foodstuff, reserved for the royal table and that of high dignitaries as well as for feeding the gods. They were an esteemed ingredient in meat and vegetable stews.

Allium cepa

Onion. *Kitāb al-Ḥashāʾish fī Hayūlā al-ʿIlāğ al-Ṭibbī*. Arabic translation of Dioscorides' *De materia medica*; revised edition of Abū ʿAbdallāh al-Nātilī (380/999) of the original Arabic translation by Ḥunayn ben Isḥāq (d. 260/873), Leiden, Universiteitsbibliotheek, MS Or. 289, fol. 92v

Onion is an annual bulbous herb in the onion family (Amaryllidaceae), the bulb formed of white succulent storage leaves. Bulb onions and slender spring onions are different forms within the same species. The leaves are green, narrow and flat. The greenish to white flowers form a rounded head at the top of the flowering stem and produce small black seeds. The origin of onions is obscure, but they may have been taken into cultivation in Central Asia, and may well have been cultivated in the Near East before their first appearance in the written record. There are no archaeobotanical records of onion seed in the Near East, doubtless reflecting the fact that seeds would have been carefully husbanded for resowing. Dried onions are found in ancient Egyptian tombs dating from 4600 BCE. Today, onion has a very wide range of traditional medical uses in the Near East. Among other uses, the bulb is taken internally to treat colds and bronchitis, headache, diabetes, and as an aphrodisiac. External uses include the treatment of wounds and sprains. The seeds are also used.

The Akkadian names for onion, *šamaškillu* or *šusikillu*, are loanwords from Sumerian *šum-sikil*, which means literally "pure bulbous plant". In western and Central Asia, where the onion possibly originated, it is known as *piyāz*, likely to be the original name. The Arabic and Hebrew word for onion, *baṣal*, alludes to its growth with many layers; the term is derived from the Arabic verb *baṣala* "to peel, to rip off". The English word onion derives from the Latin and refers to the appearance of the plant – *unio* denotes a single pearl of large size. The genus name is borrowed from Latin *alium*, "garlic", which has no accepted etymology and *cepa* "onion", a loanword from an unknown language.

Assyrian and Babylonian practitioners knew the medicinal properties of the plant, which was grown in the physic garden of Merodach-Baladan, the Babylonian king also known as Marduk-apla-iddina II.

Only one use of onion as a single ingredient is known. It was employed to treat dry eyes. The well-known lacrimatory effect of chopped onions seems to be its rationale. The medication consisted of both an internal and an external application.

"If a patient's eye is sick with dryness: chop onion, let the patient drink it with beer, instil it with sesame oil into the eye." [1]

Dioscorides (II, 151) prescribed the juice of onion for dim vision, sore throat, purulent ears, skin abrasions, dog bites, digestive problems and in the form of nasal drops to clear the head. Al-Kindī (no. 12) recommended onion for fever, jaundice and pains of the stomach and liver. Ibn al-Bayṭār (no. 296) includes a reference to the use of onion as eye drops for excessive tearing.

POMEGRANATE

Akkadian *nurmû* | Sumerian *nurma*

Punica granatum

● ● ●

Pomegranates are recorded as being cultivated in orchards in-between date palms in the earliest cuneiform texts from the middle of the 3rd millennium BCE. Administrative texts from the end of the 3rd millennium mention that the rind was used for tanning sheepskin. The pomegranate fruit formed part of the diet at the royal table and occasionally its juice was used to flavour beer. The fruit was associated with the goddess of love, the Sumerian Inanna or the Akkadian Ishtar, hence the belief that it stimulated sexual desire. In modern folklore, the abundant seeds within a fruit were an indication that it increased fertility.

The pomegranate (*Punica granatum*) is a small tree or shrub with lustrous leaves arranged opposite to each other, belonging to the Lythraceae plant family. The showy flowers are red and the petals are somewhat wrinkled. The fruit is

Fruit peel of pomegranate (*Punica granatum*), Palestine, 1924
Economic Botany Collection, Kew 54933

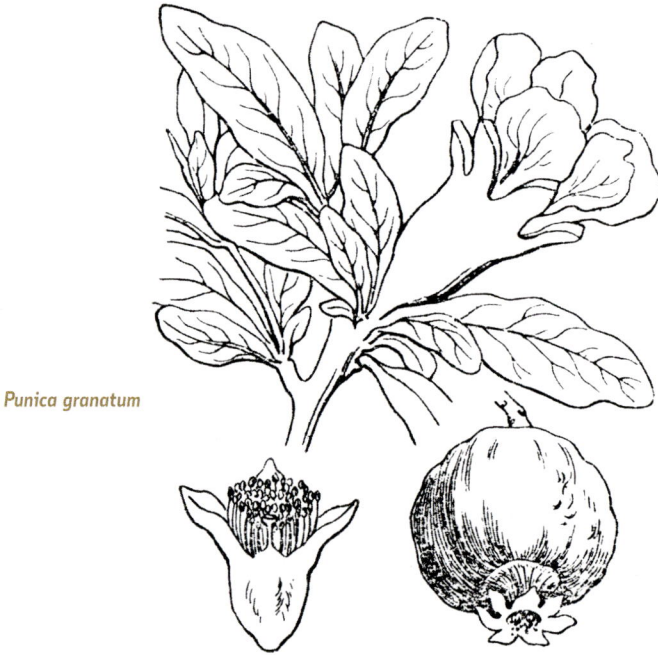

Punica granatum

spherical, up to 12cm in diameter, the rind ranging from being reddish pink, pale red to scarlet, to brownish; it is crowned with the lobes of the somewhat succulent sepals. When opened, the fruit is partitioned by thin, leathery, cream to yellow partitions, enclosing many red to pink angular fleshy seeds. Wild pomegranate grows in the region between the Balkans and the Caspian Sea; it was first taken into cultivation about 5,000 years ago. Pomegranate is common in the archaeological record, occurring at about 40 sites, including the Mesopotamian site of Nimrud (600 BCE). The presence of fruits in the Uluburun shipwreck off Turkey's southern coast (c. 1300 BCE), alongside many other luxury products, points to the special status of the fruit. Different parts of the plant have many traditional uses in the Near East today, including the fruit rind for diarrhoea, the root bark for tapeworm, the flowers as an astringent, and the fruit for stomach conditions.

The Akkadian *nurmû* or Sumerian *nurma* are wanderwords of unknown origin; similar word formations are Arabic *rummān*, Hebrew *rimmon* or Persian *anar*. The genus name comes from *punicus* meaning 'Phoenician' and 'Carthaginian' (from the city in North Africa where it was grown in Roman times) and *granatum* meaning Granada, the city in Spain into which they were exported. Its earlier name had been *malus granatum* meaning 'Granada apple' (*malus granata* translates as Grenada apple tree). The English name comes from *pomum* "fruit or fruit tree" and *granatum* "Granada". The Greek word *rhóa* for pomegranate might be linked to the verb *rhéin* "to flow" and refers to fruit juices.

Assyrian and Babylonian healers made use of almost all parts of the pomegranate: root, bark, leaves, rind, juice and fruit. Applications range from external treatments of skin diseases and sores to the administration of potions and eye, ear and nose drops.

Jaundice was understood as a condition that turned the white of the eyes yellow and as a disease of the gastrointestinal tract, and was treated with the bark, rind, root, leaves and juice of the pomegranate.

"If the patient's eyes are full of jaundice: crush bark or leaves of pomegranate, blow it with a copper tube into his eyes." [1]

"If the patient's eyes are full of jaundice: crush the rind of pomegranate with sesame oil, rub his eyes." [2]

"If the patient suffers from jaundice: crush 16.66g bark or root of pomegranate and have him drink it with beer." [3]

For open sores in the nostrils caused by the disease called in Akkadian *bu'šānu*, bark and juice were applied.

"If bu'šānu has caught the patient's nose, the nostrils ache and the nose is covered with open sores: crush the bark of pomegranate and sprinkle it over the sores, then he shall recover." [4]

"If bu'šānu has caught the patient's nostrils, instil warm pomegranate juice constantly into the nose – keep the juice warm." [5]

Bark and juice were used to treat earache caused by discharge of pus or tinnitus.

"If the patient suffers from suppurating ears: slaughter the bird called 'bird-of-the-cave', mix the bark of pomegranate with the blood, put the mixture over coals to heat it up gently, instil it into his ears." [6]

Small ivory pomegranate. The reddish colour is a stain on the surface due to long exposure to well water or sludge. The piece was possibly sewn onto a garment (c. 9th–8th century BCE)
© The Metropolitan Museum of Art, New York, 54.117.7, Rogers Fund, 1954

"If pus is dripping from the ears of a patient: crush dried pomegranate, blow it into his ears or alternatively drop pomegranate juice into his ears." [7]

"If the patient's ear is ringing, mix cedar tar with pomegranate juice, instil it into the ear, then he shall recover." [8]

The ground-up rind and bark were helpful for various skin diseases.

"If the patient is covered with skin eruptions, crush pomegranate rind, mix it with ghee and apply it to the sores." [9]

"If a skin eruption appears on the patient's foot, ripens like an abscess, and exudates, the patient shall recover. Crush dried pomegranate rind, sprinkle it over the sore." [10]

"If the patient suffers from a gangrene of the foot: roast pomegranate bark, crush it, rub the foot with fish oil and sprinkle the bark over the foot." [11]

"If the patient suffers from a gangrene of the foot: roast pomegranate rind, apply it to the sores." [12]

Sculptured relief panel from Sennacherib's Southwest Palace at Nineveh showing men carrying grapes and pomegranates (c. 704–681 BCE)

© The British Museum, BM 124799

Pomegranate. *Kitāb al-Ḥ ashā'ish fī Hayūlā al-'Ilāğ al-Ṭibbī*. Arabic translation of Dioscorides' *De materia medica*; revised edition of Abū 'Abdallāh al-Nātilī (380/999) of the original Arabic translation by Ḥunayn ben Isḥāq (d. 260/873), Leiden, Universiteitsbibliotheek, MS Or. 289, fol. 45r

© Leiden, Universiteitsbibliotheek, CC BY 4.0

Pomegranate-shaped pendants of a necklace found in an Assyrian tomb. Tomb no. 45, field no. Ass 14630, c. 14th–13th centuries BCE, Ashur

© Staatliche Museen zu Berlin, Vorderasiatisches Museum, VA Ass 4346 and VA Ass 4347, Photo: Olaf M. Teßmer

Wild pomegranate flowering.
*Kitāb al-Ḥashāʾish fī Hayūlā
al-ʿIlāğ al-Ṭibbī.* Arabic
translation of Dioscorides'
De materia medica; revised
edition of Abū ʿAbdallāh
al-Nātilī (380/999)
of the original Arabic
translation by Ḥunayn ben
Isḥāq (d. 260/873), Leiden,
Universiteitsbibliotheek, MS
Or. 289, fol. 45v
© Leiden, Universiteitsbibliotheek,
CC BY 4.0

A poultice was made to alleviate pains located in the groin.

"If the patient's groin is hurting: crush pomegranate rind, apply it on the aching part." [13]

The fruit and the juice were suitable for the stomach.

"If the patient suffers from colicky pain in the abdomen: let him drink pomegranate juice or eat pomegranate." [14]

The leaves were used to remove worms and tapeworms.

"If the patient suffers from intestinal worms: choose the leaves of a pomegranate that grow on the northern side of the tree, add them to water and enclose it in a kiln, remove it the next morning, filter it, have the patient drink it for three days, have him vomit, then the worms will be expelled." [15]

Too much bleeding during menstruation was stopped by introducing a wad of absorbent material containing pomegranate bark.

"If a woman suffers from bleeding: crush pomegranate bark, wrap it in wadding (and apply it)." [16]

Practitioners ascribed aphrodisiac properties to the juice of pomegranate. To stimulate women, the following ritual was to be performed:

"Recite three times the incantation 'A beautiful woman has come to make love with me, the goddess Inanna who loves apples and pomegranates has come to stimulate me. Get erected, ejaculate, be a stone, get straight and get erected again! It is all Inanna's doing – she watches over matters of sex and love.' Give the pomegranate to the woman, have her suck its juices, then the woman is ready to come to you and you can make love with her." [17]

Dioscorides (I, 110) recommends pomegranate for gastrointestinal problems, to treat women's periods, sores in the mouth and genitals, afflictions of the nostrils and earache. A poultice was helpful for intestinal hernias. The flowers of pomegranate were used for the same diseases. Al-Kindī (no. 65) uses the flowers in poultices for problems of the liver, stomach and spleen, to strengthen the limbs and to treat throat pain, abscesses and decayed teeth.

POPLAR

Akkadian *ṣarbatu* | Sumerian *giš-asal₂*
Populus euphratica or *P. alba*

● ● ●

Poplar is one of the few trees that figure in the mythology of ancient Mesopotamia. The tree grows in the sanctuary of the weather god Adad to give shade. The tale of the legendary king Etana begins with a fable of a serpent and an eagle that live peacefully together in the roots and crown of a poplar, until one day the eagle devours the serpent's brood. The connection between the tree and the divine world is visible in another tale, about the gardener Shukaletuda and the goddess Inanna. The goddess, exhausted from her travels, sleeps under the wide branches of the poplar that grew in her holy garden. One day the gardener sexually assaults the goddess: in her rage, Inanna brings plagues over the land in revenge.

The motif of poplar as a shade tree was deeply anchored in daily life – kings would praise themselves for planting tall and broad poplars in towns and gardens.

Populus euphratica
Reproduced with permission from *Flora of Iraq* 4(1): 29, pl. 6

Wreath of poplar leaves strung with beads. Jewellery associated with the body of Puabi found at the "Royal Cemetery" of Ur (tomb RT 800, c. 2600–2450 BCE). Hammered gold leaves strung with laps lazuli cylindrical and carnelian truncated bi-conical and lentoid beads
© University of Pennsylvania Museum of Archaeology and Anthropology, B17710 (U.10935*bis*)

The poplar formed part of the riverine landscape of Mesopotamia and was used as a place name. The tree stimulated the ancient mind to tales of a god "Lord-Poplar" and a goddess "Lady-Poplar". The popularity of the tree turned the leaves, imitated in hammered gold, into the adornment of the mid-3rd-millennium BCE queen Puabi.

The poplar – alongside the tamarisk – was the standard wood of roof beams, furniture and good-quality fuel and was cultivated in gardens and forests at least from the middle of the 3rd millennium BCE onwards. Cuneiform documents mention trees growing up to 18m high. Two species of *Populus*, species of tree in the willow family (Salicaceae), are relevant to a discussion of the medicinal use of poplar in Mesopotamia. The Euphrates poplar (*Populus euphratica*) is a small tree growing up to 15m high. The spreading branches provide shade, so it is planted along roads but today it is not recommended for gardens as it produces root suckers, leading to thickets. Its distribution spans the region between northwest Africa and Mongolia and the western Himalayas. In Iraq, it is common in the lower forest zone and lower hills, growing by mountain streams and riverbanks, along canals and in ditches. Because of its quick growth, it is often coppiced; the leaves make good fodder and the branches are used for fuel, and for fences and other utilitarian objects. The timber has a long history of use as poles for rafters. A second species, *P. alba*, a native of Europe, the Near East and Central Asia, is cultivated in the lower mountain valleys of northern Iraq, but also grows spontaneously along streams, spreading by root suckers. This is a tall, graceful tree up to 30m tall, with greyish-white bark and a broad crown, sometimes becoming pyramidal in shape. Poplar charcoal (of undetermined species) is found at many archaeological sites, reflecting its usefulness as a renewable source of firewood and its use in construction.

The meanings of the Akkadian and Sumerian terms for the poplar, ṣarbatu and giš-asal₂, respectively, are unknown. The Arabic terms *gharab* and Hebrew *'arabah* are related to the Akkadian name. The genus name *Populus* is based on the Latin *arbor populi* ("tree of the people"), as it was so widely planted in public places during Roman times; the species name *euphratica* refers to its habitat growing along the river Euphrates; *alba* "white" refers to the whitish underside of the leaves of *P. alba*.

Healing practitioners used the ashes and the resin, which they called "white substance" or "white drug"; by contrast, the leaves were rarely prescribed, and only in compound medicines. One of the oldest preserved recipes, written in Sumerian and dating to the end of the 3rd millennium BCE, mentions the use of the ash or charcoal. It is not always possible to differentiate between ash or charcoal because both plant products are written with the same cuneiform sign. About 1,500 years later, the charcoal was used in a wad of absorbent material, possibly to stop bleeding during pregnancy.

"If a woman suffers from problems during pregnancy, crush coal of poplar wood, wrap it into fleece (and introduce it into her vagina)." [1]

Practitioners employed the resin or sap when it was hardened. Only one recipe recommends the fresh resin for treating a rectal disease.

"If a man defecates and evacuates either mucus, pus or bloody excrement from his sick anus, in order to heal him: boil fresh 'white drug' in milk and (have him drink it)." [2]

The resin was taken internally as a potion in case of severe digestive problems, violent cough and diphtheria; as a salve, it was employed in the treatment of skin eruptions and fever, and as a daub for afflictions of the eye.

"If the patient is sick with carbuncles, rub him first with ghee, then crush 'white drug' and sprinkle it over the affected skin." [3]

"Drug to stop skin eruptions that affect the head and eyes – 'white drug' is to be crushed and rubbed with sesame oil." [4]

"If red, itching and weeping eczema appear on the patient's feet, in order to heal him: crush 'white drug', apply it on the skin sore." [5]

"If the patient's eyes are bloodshot, pound 0.23 g resin of poplar with sesame oil, daub the eyes." [6]

لوقا وهو الجوز وسم السكان
بالعراق وجوزبستان سمي الجبع ١
وقشرهن النجه اذا اسر منه وزمانا نفع من عرق النسا وقطين البول
وقديقال انه بنفع المراة
الحبلا خاولا اذا اراد اسر بذمنه
وكذلك بفعل اذا اسرب ورقه
مع حلب عطرطم هاو عصارة
الور واذا قطر بالاد رو هو فار
نفع من الطها وثمر الجوز اذا اخذ
منه الدا خرج بسع ورقه حبه
ادادو وخلط بعسلا واكبليه
ابراغتناوه العبر وقد برم رقوم
ارا لجبر الروم وغيره اذا قطع
صعاراد وا وغزبر ع ارض سخه
ما لحه اثبت السنه كلها
قطبا بونكل ٠

ما قبر وسمه اهل النثام الدار كسنه
وعم توم انه السبانه ومعم اخزور انه الطراست وهو فشور بوني يه من
اذصر وبر لونه الي الشعره ما هو علط فابض حرا و فردشرب لنفت الدم و
وجه٢ معاوسلار الفصول الي البطن
اعبروس وهو الجوز الروى
ورواع روس وسبع سبع مروح الفرس
وصرنافاذا وضع عله خاريصمه
فربيع واخلاط المراهم وبدبعال الركبه
اذا اسرب خارج مع الصبح وبهال
اضصال الدرسل بص صبعه والنهر
الدرسم ارد اوبس خدبع النهر وصا

Poplar bark (*Populus euphratica*), Karachi. Pakistan, 1886
Economic Botany Collection, Kew 41353

Occasionally, the emetic effect of the potion is stated. Whether this effect can be extrapolated to all potions remains open.

"If the patient's nose is feverish hot and he salivates, the bu'šānu *disease has caught him: crush 'white drug', have the patient drink it with beer so that he vomits, then the patient will recover."* [7]

"If the patient suffers from violent bouts of cough spitting bloody sputum, crush 'white drug', fold it into a mixture of high-quality sesame oil, syrup, and first-quality beer, hold his tongue, feed in the liquid drop by drop until the patient vomits." [8]

"If the patient's belly is sick, have him drink 'white drug' with sesame oil, then he shall vomit." [9]

"If the patient's belly is sick and he has piercing pain and does not eat or drink, have him drink 'white drug' with sesame oil." [10]

Fever was treated with salves and poultices.

"If a patient's head is feverish hot, in order to remove the fever: prepare a mass of poplar (resin) and mustard water, apply it as a bandage." [11]

"If a patient has fever due to heat stroke, crush 'white drug', mix it with sesame oil, anoint the patient repeatedly, then he shall recover." [12]

In the case of urinary retention or pus coming from the ear, a small copper or reed tube was used to introduce the medication.

"If a man holds back something hard when urinating, crush 'white drug', mix it with sesame oil, blow it with a copper tube into his penis." [13]

"If the patient's ears discharge pus, crush 'white drug', blow it with the help of a small reed tube into his ears." [14]

Assyriologists agree that the Sumerian *ğiš-asal₂* and Akkadian *ṣarbatu* refer to the Euphrates poplar. However, Iraqi traditional medicine uses the bark and leaves of white poplar as a diuretic, stimulant, tonic, anti-rheumatic and febrifuge; Euphrates poplar is not included in today's herbal pharmacopoeia.

Dioscorides (I, 81) recommended bark, leaves and catkins of the white poplar. The bark was used for hip diseases and afflictions of the urinary tract; the leaves caused sterility, their juice served as a treatment of aching ears; and the catkins helped in case of dim-sightedness. Al-Kindī (no. 85) lists the resin of poplar as a single ingredient and the bark was prepared to treat leprosy. The modern and ancient uses compare well with the cuneiform evidence.

Note, however, that the Euphrates poplar forms part of the pharmacopoeia of traditional medicine in Jordan; it is used as a febrifuge, diuretic, stimulant and anti-rheumatic. A tincture of the catkins is applied to skin diseases and a maceration of the catkins is used for bronchitis, arthritis, rheumatism and pneumonia. The coal made from the wood is a carminative and anti-flatulent.

RED BRYONY

Akkadian *imḫur-līm* "It-faces-a-thousand"

Bryonia multiflora

● ● ○

Red bryony is a wild plant used to treat many illnesses in ancient Mesopotamia. Assyrian and Babylonian healers described it with the following words: "*The plant creeps over the ground like colocynth, its tendrils look like that of a melon, its leaves are divided, its seeds look like that of madder, its root is bitter and soft. Its name is 'it-faces-a-thousand'.*" [1]

Bryonia multiflora is a plant in the cucumber family (Cucurbitaceae) with weak stems that can scramble over trees and shrubs, or lie prostrate on the ground. It has a tuberous root. The leaves have simple tendrils, and the leaf blade is deeply 5-lobed with the lobes sometimes lobed in turn. The flowers are yellowish-green with the male and female flowers found on separate plants. The fruit is a round berry, 7–8mm in diameter, that becomes red and shiny when ripe. Red bryony is native to Turkey, southern Syria, northern Iraq and west and

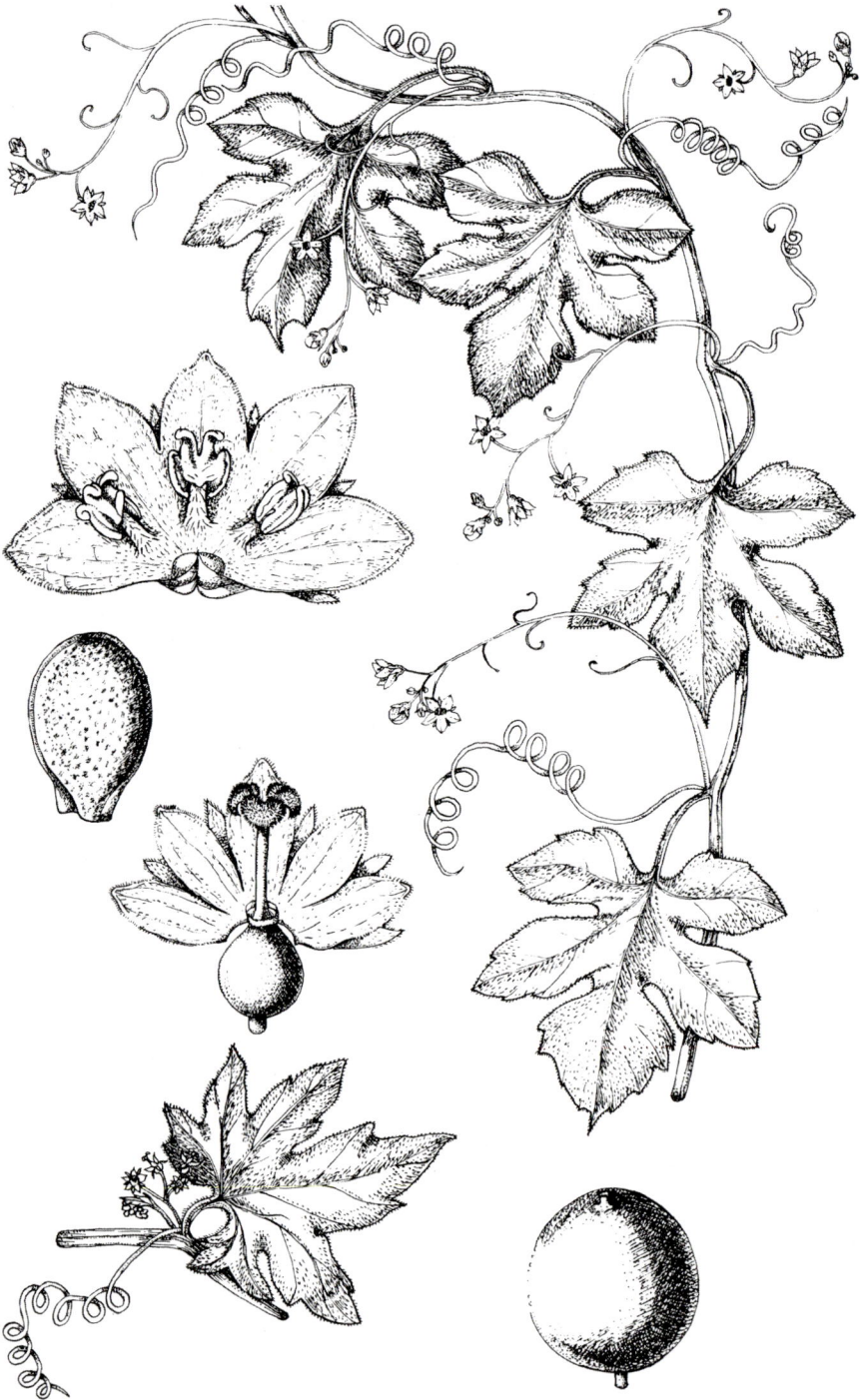

Bryonia multiflora
Reproduced with permission from *Flora of Iraq* 4(1): 201, pl. 38

southwest Iran. It is the only *Bryonia* species to occur in Iraq, where it is found on rocky slopes in the northern mountains, flowering from April to June. Seeds have been found at just a few archaeological sites.

Red bryony, and white bryony (*B. alba*), are mentioned in the earliest Arabic pharmacopoeia, *al Aqrābādhin al-saghīr* (*Dispensatorium Parvum*) written by Sābūr ibn Sahl (a Nestorian physician and pharmacologist, d. 869 CE). The main uses listed are as a diuretic and laxative and for cleansing. The tuberous roots were also listed as a treatment for diabetes. Chakravarty & Jeffery, writing in *Flora of Iraq*, note that there is no record of the use of this plant in present-day Iraq, but it is likely to have been used in earlier times as a medicinal herb. Al-Rawi (1964) lists *Bryonia dioica* (now *Bryonia cretica* subsp. *dioica*) in *Medicinal Plants of Iraq*, but this species is not recorded for Iraq; this record probably refers to *B. multiflora*.

The Akkadian name *imḫur-līm* "it-faces-a-thousand" follows the pattern of the plant name *imḫur-ašra* "it-faces-twenty" (black bryony, an unrelated species, see page 63), both referring to the many diseases for which they were used. The botanical term from which the English name and genus name are derived goes back to Greek *bruéin* "to sprout abundantly"; the species name *multiflora* refers to the plant's "many flowers". Ibn al-Bayṭār (no. 1654) gives as the Arabic name *fāširā* meaning "dissolver", possibly referring to the strong laxative properties of the root.

Red bryony was employed for a broad spectrum of diseases. It was administered externally as a salve and rubbed into the skin, and taken internally as a potion. No reference to the plant parts that are used is given; practitioners probably used the tuberous root. Occasionally, the required amount is given, suggesting that the toxicity of the plant was well-known. As a diuretic and emetic, a potion served to treat stricture of the urethra and afflictions of the gastrointestinal tract. Red bryony was also recommended for infertility and problems during pregnancy. Practitioners made a salve with sesame oil that was helpful for chest diseases, animal bites, skin sores and various dermatological diseases; they considered "it-faces-a-thousand" one of the principal plants for jaundice and witchcraft.

Only exceptionally is the emetic effect of treating biliary diseases stated; usually recipes do not include references to the desired effect of the medicine. *"If the patient's bile is affected, crush 'it-faces-a-thousand', have him drink it with beer, then he shall vomit."* [2]
"If the patient suffers from sick bile, have him drink 0.7g 'it-faces-a-thousand' with half a litre of sesame oil and beer." [3]
"If the patient's body is yellow, his face is yellow, he has lost weight. The name of the disease is jaundice. Crush 1.3g 'it-faces-a-thousand', have him drink it with beer." [4]

Severe abdominal pain located in the stomach and the belly was usually treated with a potion: in some cases, the mode of consumption was prescribed.

"If the patient suffers from colic, he should eat 'it-faces-a-thousand' on an empty stomach." [5]

"If the patient's stomach is sick, crush 'it-faces-a-thousand', mix it with first-quality beer, have the patient drink it on an empty stomach, then he shall recover." [6]

"If the patient constantly feels a crushing pain in the belly, have him eat 'it-faces-a-thousand' on an empty stomach, drink it with beer, rub him with sesame oil, wrap it in fleece." [7]

The plant was given as an enema to expel intestinal worms; the following recipes belong to the few examples that give the exact amount of the drug.

"If the patient's belly is infested with intestinal worms, tapeworms or their eggs, crush 0.65g 'it-faces-a-thousand', pound it together with filtered sesame oil, introduce it into the anus, then the patient shall recover." [8]

That beneficial function on bladder function was ascribed to red bryony can be inferred from its use in treating stricture of the urethra.

"If a man holds back something hard when urinating, crush 'it-faces-a-thousand', have him drink it with beer." [9]

It was said about the plant that it is "good for almost all skin sores". [10]

Indeed, a variety of prescriptions deal with treatment of the skin.

"'It-faces-a-thousand' – drug for serpent bite – to be crushed and rubbed with sesame oil, to be sprinkled over the bite, to be drunk with beer." [11]

"'It-faces-a-thousand' – drug for scorpion sting – to be drunk with first-quality beer, to be rubbed constantly with sesame oil, to be fumigated like incense." [12]

"If the patient suffers from itching skin eruptions on the head, crush 'it-faces-a-thousand', rub the sore skin." [13]

A salve was recommended for a chest disease and a daub was used for blurry eyes.

"'It-faces-a-thousand' – drug for blurred vision." [14]

The potion was considered useful to stop bleeding during pregnancy and to treat women who could not give birth.

"A woman suffers from bleeding during pregnancy, crush 'it-faces-a-thousand', have her drink it with first-quality beer." [15]

"'It-faces-a-thousand' – drug for a woman who cannot give birth – to be drunk with beer." [16]

The role of red bryony in magical rituals is well described in the elaborate ceremony to counterattack the assault of witches and warlocks called "burning",

Maqlû. The Akkadian title refers to one of the rites, the burning of effigies of the evildoers in fire. To release from witchcraft and to protect the bewitched person, the following was to be done:

"Spin a cord of combed wool, knot seven knots; knot into each knot red bryony. Recite the incantation 'You, "it-faces-a-thousand", are a plant known from time immemorial, that undoes anything, whose tip reaches the sky and whose roots fill the earth. The witch saw you and she turned pale, her lips became dark. You are the plant that does not allow evil words or hate-magic.' Recite the incantations seven times over the cord, hang it around the person's neck, then witchcraft shall not approach him." [17]

Another magic spell of the same ceremony conjures the plant with the words *"May 'it-faces-a-thousand' slap the witch's cheek!"* [18] This wish might actually represent the skin-irritating effect of the root. The fresh sap initially causes redness of the skin that can become a painful inflammation with blistering. The treatment of skin sores probably goes back to this effect. Eating red bryony was another way to treat bewitched people.

"'It-faces-a-thousand' – drug against witchcraft – to be eaten at New Moon." [19]

Dioscorides (IV, 182 and 183) refers to other bryony species, which are identified as *Bryonia dioica* and *Bryonia alba* or *Tamus communis*. For the debate on whether the ancient Greek name *ampelos melaina* "black vine" refers to *Tamus communis* or *Bryonia alba*, see Renner *et al.* 2008. Ibn al-Bayṭār (no. 1654) refers to the emetic effect of the juice of red bryony.

SESAME

Akkadian *šamaššammu, šamšammu* "Oily plant" | Sumerian *še-giš-i₃* "Oil-containing plant"

Sesamum indicum

● ● ●

Sesame was the second most important field crop after barley in ancient Mesopotamia. It was cultivated from at least around 2300 BCE onwards, possibly introduced through Mesopotamian sites that were in trading contact with India, where sesame was first cultivated. Farmers differentiated between an early-planted sesame, sown at the beginning of the year, and normal sesame,

Sesamum indicum
Reproduced with permission from *Flora of Iraq* 4(2): 642, pl. 116

sown in late summer. Sesame oil had a wide range of applications: it was used to wash wool textiles, fabricate soaps, grease leather, treat wooden planks for boats, as a base for perfumes and as a fuel for lamps. It was used in cooking and baking bread and cakes. Typical cakes would be made of barley flour, dates and sesame oil, to which fresh and dried sesame cake or raisins could be added. However, sesame oil was not easily digested so its consumption became proverbial in ancient Mesopotamia: "farting like someone who has eaten sesame oil". The oil was also used in body care. Soldiers and travellers would protect themselves from the cold by anointing their feet with the oil. Sesame was so ubiquitous in ancient Mesopotamian daily life that the Greek historian Herodotus reports in his *Histories* (Book I, 193) that Babylonians "use no oil except what they make from sesame".

Sesame is a small annual or perennial herb in the Pedaliaceae plant family, widely cultivated as a summer crop in Iraq. The flowers are white, purplish or reddish, producing an oblong capsule bearing numerous small seeds. Archaeological evidence indicates that sesame as a crop originated on the Indian subcontinent and spread to Mesopotamia, Egypt and China. For many years, sesame seeds were puzzlingly absent from the archaeobotanical record of the Near East, including Mesopotamia. Experiments showed that when burnt, the seeds of sesame are far more fragile than those of flax. Careful recovery of ancient plant remains has led to seeds being found at about ten sites, including the Iraqi site of Abu Salabikh (3200–2300 BCE). Sesame oil, *tahini* and sesame seed garnish on bread, cakes and sweetmeats is popular and relished in Iraq and western Asia. Traditional uses of sesame in the Near East include use of the oil to treat burns, and the seeds as a laxative and to treat bronchitis, coughs and other chest conditions.

The Akkadian name for sesame, *šamaššammu* or *šamšammu*, is the result of the bound combination of the words *šamnu* "oil" and *šammu* "plant". The Sumerian term *še-giš-i₃* has a similar meaning, namely "oil-containing plant". From Akkadian, the plant name has entered the vocabulary of other Semitic languages such as the Arabic (*simsim*) and Hebrew (*šumšum*), as well as many other languages such as the English sesame.

Sesame oil was one of the basic carrier substances in medicine; it was used for preparing salves, bandages, poultices and potions. Occasionally it was prescribed as a preparatory step during multiphase treatments. To treat various afflictions of the eyes, described simply as "sick eyes" or "dryness", sesame oil was applied either to the temples or directly on the eye.

Sesame seeds
(*Sesamum indicum*),
India, 1926
Economic Botany
Collection, Kew 46408

"If a patient's eye is sick, drip about 10g sesame oil on his temples. Spread copper patina from a leatherworker on a leather strip, bind it around the temples." [1]

This is followed by the prescription for an eye daub.

"If a patient's eye is sick with dryness: chop onion, let the patient drink it with beer, instil sesame oil into the eye." [2]

Sesame oil does not seem to have been used as a single ingredient in medicine, it rather appears in combination with other substances. The following recipe recommends a mixture of the oil with ghee, again for treating the visual organs.

"If a patient suffers from night blindness, mix ghee and first-quality sesame oil, daub the patient's eyes repeatedly." [3]

There were sesame oil experts in ancient Mesopotamia, called in Akkadian *ṣāḫitu*, "preparer of sesame oil", after one of the two extraction methods to produce the oil: first the seeds were roasted in an oven, then ground with a millstone into a pulp, which was then mixed with water and heated in order to extract the oil. The residual sesame cake received a specific name in Akkadian, *kuspu* (or *kupsu*), and is known in the modern Middle East as *kusup* in Iraq or *ksebe* in Syria, where it is used as food for cattle. The Sumerian term is *duḫ-še-giš-i₃*. The other method involved pounding the sesame seeds in a wooden mortar. This process could be repeated at least six times and resulted in a high-quality oil.

Practitioners made use of the *kuspu* sesame cake: it was dried and pounded into flour, which served as the carrier substance for poultices and bandages. In order to treat gangrene of the feet, as well as to reduce fever and afflictions of the vision, it was employed as a single ingredient.

"If a patient suffers from gangrene of the feet, crush sesame oil cake, gently rub and daub the skin sore with it." [4]

"If a patient's head is feverish hot, the temples are throbbing, the eyes are affected – be it that the vision is blurred, cloudy or impaired, be it that the patient suffers from double vision, red and sore eyes – so that the patient constantly sheds tears, pound 83g of dried sesame oil cake into powder, sift it, prepare a poultice with mustard water, bandage the patient." [5]

Religious experts used the sesame cake as a raw material in some of their rituals. During the highly complex ceremony against witchcraft that began at sunset and lasted until the next morning, the expert manufactured an effigy of the witch using sesame cake. He then burnt the effigy and in doing so removed their harmful doings. Witches were believed to cause severe physical and psychological afflictions.

Dioscorides (II, 99) warns that sesame is bad for the digestion. He recommends that the plant is used as a poultice to treat inflammations, burns, ear afflictions and pains of the colon. As salve, it is helpful for headaches, and boiled with wine, it is especially well-suited for inflammation of the eyes. Al-Kindī (no. 152) recommends sesame for ear inflammation and the oil to remove abscesses, to alleviate toothache, and for cough and numbness.

SWEET FLAG

Akkadian *qānu ṭābu* "Sweet reed" | Sumerian *gi-du₁₀* "Sweet reed"

Acorus calamus

● ● ○

Sweet flag is a semi-aquatic plant that grows on the edges of lakes, rivers and swamps. Its natural habitat is Asia (Central Asia to China and the Far East) and North America. It has been introduced into Europe, Asian Turkey

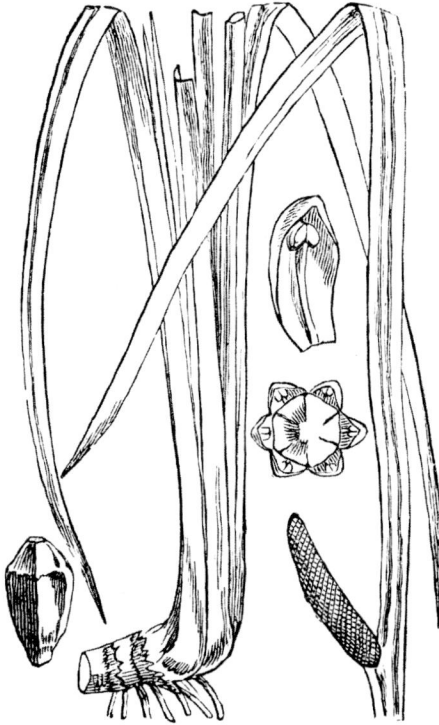

Acorus calamus

and the Caucasus. Sweet flag (*Acorus calamus*), in the Acoraceae family, has a thick, cylindrical rhizome that runs horizontally across the wet ground, from which the roots or rootlets arise. The rhizomes emit a nutty, leathery and cinnamon-like scent. The leaves grow up to a metre long, are narrow, sword-shaped and flat, and when crushed emit a tangerine- or citrus-like smell. The numerous flowers are borne on a hard stem-like structure (spadix) and are sweetly fragrant. Because of its warm, spicy and woody notes, the essential oil of sweet flag rhizomes (and sometimes leaves) is a common ingredient in the perfume industry – not only nowadays but also in ancient Mesopotamia and ancient Egypt. Sweet flag is, however, absent from the archaeobotanical record. Even in the unlikely event that pieces of rhizome were to become charred, such root fragments present considerable challenges to identification under the microscope.

According to the 1st-century CE Roman naturalist Pliny the Elder (XII, 48), *Acorus calamus* grows in Arabia, India and Syria, the latter being the best. And indeed, the city of Mari, located in today's Syria, was one of the ancient Mesopotamian perfume centres where sweet flag was processed in great quantities. Syria lies outside the wild distribution of the plant, so it was probably cultivated or imported. The Akkadian, Sumerian and English names of the plant originate in the sweetish smell of the root (Akkadian *ṭābu* and Sumerian *du₁₀-ga* mean "sweet"). The Latin term *calamus* comes from ancient Greek *kalamos*, which like Akkadian *qānu* and Sumerian *gi* means "reed". Latin *acorus* goes back to Greek *ákoron*, which the ancient Greeks derived from the term *kórē* "pupil of the eye" because the plant was used for eye care. The common English name, sweet flag, refers to its sweet smell and similarity to *Iris* species, commonly called "flags".

Sweet flag was traded into ancient Mesopotamia as early as the first half of the 3rd millennium BCE. Merchants in Sumer would receive between 1kg and 5kg or exceptionally about 28kg of sweet flag, possibly carried by animal caravans coming from Central Asia. It could also have been imported via the sea trade from Arabia and India, both important trading partners of Sumer during the 3rd millennium BCE. The dried rhizomes were used to scent cosmetic salves and ointments. In later times, around the beginning of the 2nd millennium BCE, even larger amounts were recorded, with sweet flag replacing juniper as the most used aromatic.

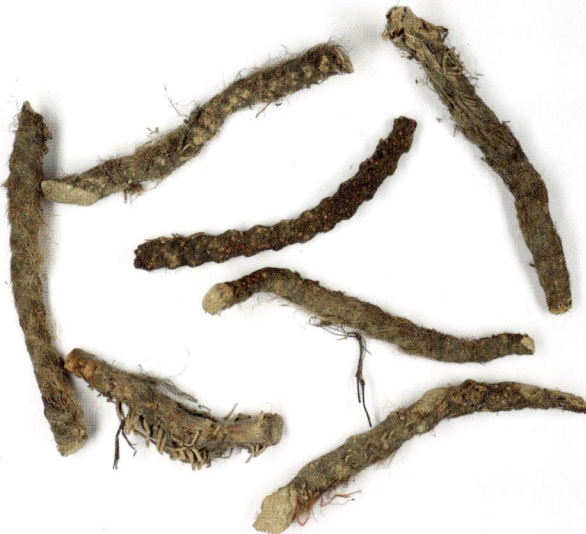

Rhizomes of sweet flag (*Acorus calamus*), India
Economic Botany Collection, Kew 34513

A few Akkadian recipes dating from the middle of the 2nd millennium BCE provide insight into the making of fragrant ointments. One of the basic ingredients was sweet flag. Flower, leaves and rhizome were carefully selected, crushed and mixed with boiling water. Further aromatics could be added. The mixture was then left overnight to macerate and strained on the following morning. The liquid, to which again sweet flag and other aromatics could be added, was heated and mixed with oil. The mixture was left for three to four days, heated again to evaporate water, cooled, strained, placed in a small flask and was ready for use as a scented salve.

Sweet flag was used in cosmetics, and as a fumigant in incense mixtures; it also served to enhance the taste of beer, especially of emmer wheat beer. It is not always readily clear whether the fresh or the dried plant was used in the preparation of beer. Medical practitioners, however, seemed to have used dried sweet flag, which could be stored and was available in Assyrian stores.

Sweet flag. Dioscorides Pedanius of Anazarbus, *De materia medica*. Constantinople, c. 512 CE. Cod. Med Gr. 1 (facs.), fol. 58v
© Österreichische Nationalbibliothek, Vienna

Assyrian and Babylonian practitioners would drip the juice of sweet flag into the ears of a patient to treat noises in the ear. Buzzing or ringing in the ears was associated with an attack by the frightening 'spirit-of-the-dead', a demon that would enter in the patient's body by their ears, causing on its way through the body severe pains in the neck, headaches, abdominal pain and paralysis.

"If the hand of the 'spirit-of-the-dead' has caught a patient and his ears are roaring: crush sweet flag with sesame oil, apply it to his ears." [1]

Both the young green shoot and the acrid and pungent juice of sweet flag were processed into a salve for numbness and pains in the muscles.

"Acrid juice of (sweet) reed – drug for treating numbness – to be crushed and anointed with sesame oil." [2]

"Shoot of (sweet) reed – drug for treating numbness – to be crushed and anointed with sesame oil." [3]

The leaves of sweet flag, once finely ground and sifted, were drunk to treat the symptoms of the bite of a particular snake.

"Leaves of (sweet) reed – drug for the 'winged snake'." [4]

Another remedy was thought to bring relief if a woman had difficulty giving birth.

"Pour sweet reed and sesame oil into a (hollow) reed. Rub it gently over her belly, then the baby should come out straight away." [5]

TAMARISK

Akkadian *bīnu* | Sumerian *šinig*

Tamarix species, including *Tamarix aphylla*

● ● ●

In ancient Mesopotamia, tamarisk was appreciated for its shade, as a wind break and for stabilising the banks of canals. Trees were often planted along the edges of gardens and fields and on dikes surrounding farmland. Tamarisk had many medical and magical applications.

Tamarisk (*Tamarix* spp.), in the saltcedar family (Tamaricaceae), grows as evergreen shrubs or trees, sometimes forming thickets. Several species are desert trees, growing on sand dunes and tolerating saline soils: other species

Galls of tamarisk (*Tamarix gallica*), Madras, India Museum (before 1879)
Economic Botany Collection, Kew 66760

grow along streams and riverbanks and in coastal areas. Tamarisk trees have long taproots that can reach water tables at a depth of 30m; the superficial side roots can be 50m long. Branches appear leafless, as the leaves are small, green and scale-like. The leaves and young shoots bear salt-excreting glands, hence one of its common names is "salt cedar". The flowers are small, pink to white in large sprays, and the minute seeds are adorned with a tuft or crest of hairs. Several species of scale insects live on tamarisk, consuming the sap of the tree and exuding a sticky sweet liquid that crystallises as it dries. The exudate of scale insects has been traditionally used for sweetmeats, and medicinally as a laxative. It is the most convincing candidate for biblical manna, being one of the trees that grows in the Sinai desert. Tamarisk also produces gallnuts – tannin-rich growths – in response to attack by insects.

There are about 54 species of tamarisk growing in arid and semi-arid regions, distributed from the Mediterranean region to southern Africa, and from Arabia through Iran, Pakistan to Central Asia. In modern Iraq, eleven species of tamarisk are recorded. Cuneiform descriptions praising the tamarisk as a shade-providing tree, and administrative records of the delivery of trunks up to 7m long, suggest that the Akkadian and Sumerian terms refer to *Tamarix aphylla* ("leafless tamarisk"), the tallest of the Iraqi species, with

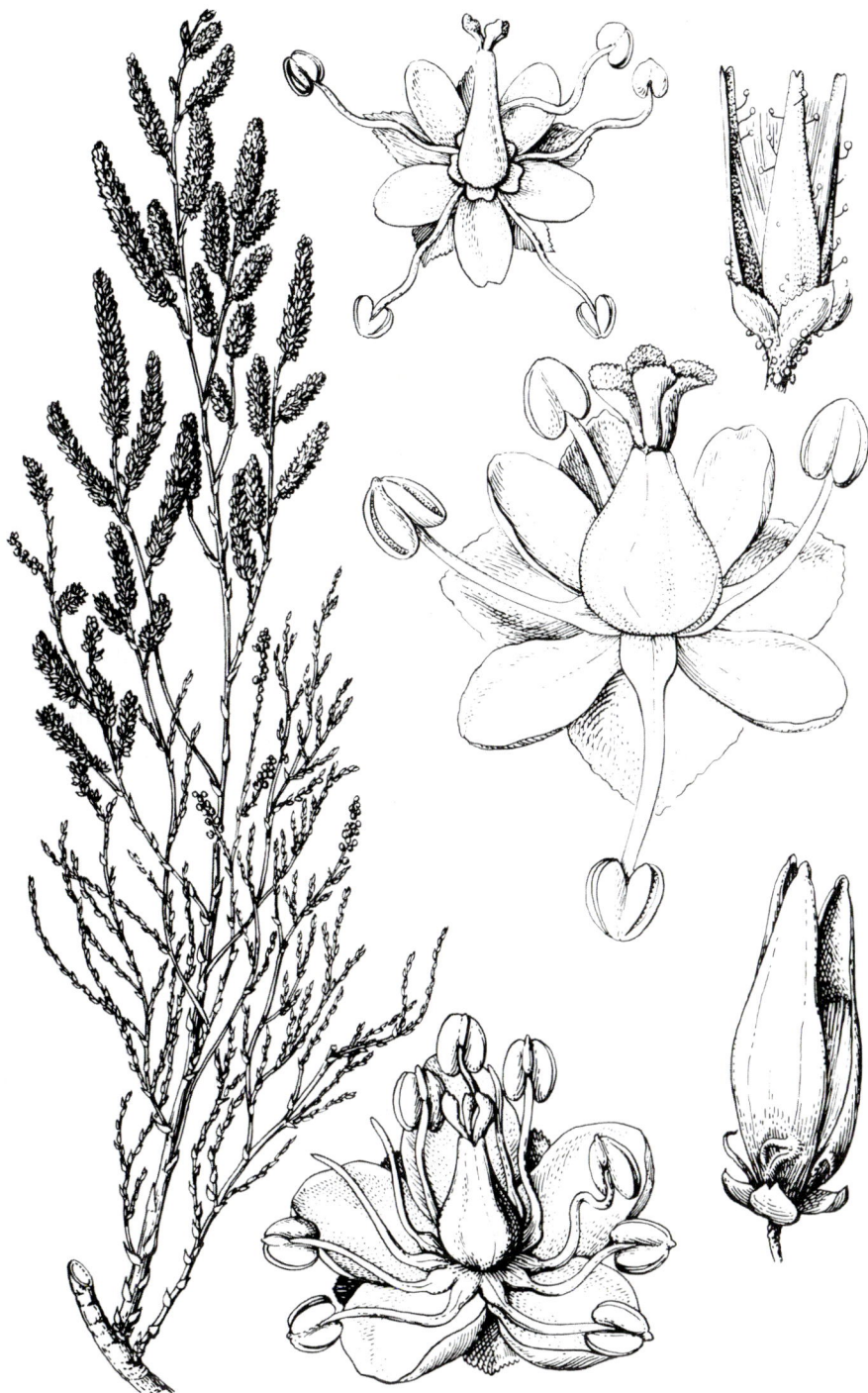

Tamarix species
Reproduced with permission from *Flora of Iraq* 4(1): 163, pl. 31

trees recorded in the twentieth century as up to 24m high. Although *T. aphylla* is today only found in cultivation in Iraq, it may previously have grown wild there. Other species grow wild in abundance in mixed shrubland and desert regions, and were no doubt harvested for medicinal use.

The remains of *Tamarix* wood, gathered as fuel, have been found at over 40 excavations in the ancient Near East, including the Mesopotamian cities of Larsa and Sippar, but the wood cannot be identified to species. In traditional medicine in the Near East, the leaves, roots and bark of *Tamarix* are used as an appetiser, diuretic and tonic, for jaundice, wounds and abscesses, and for fever and eye diseases. The gallnuts, bark and twigs contain high levels of tannin, which acts as a hepatic and stomachic stimulant to aid the liver functions and to increase the appetite. Because of its anti-inflammatory properties, tamarisk is said to promote wound healing.

The meanings of the Akkadian name *bīnu* and the Sumerian name *šinig* are unknown. The Arabic name, *aṯl*, refers to the characteristic plant root and is derived from the Arabic word *aṯala* "to be deep-rooted". In Iraq, the tamarisk is called *ṭarfa*, which is a straight loan from the Akkadian *ṭarpu'u*, another name for the tamarisk. The English name, tamarisk, comes from Latin *tamarix*, which is derived from the Sanskrit term *tamalaka*, referring to a tree with a dark bark. The Akkadian term for gallnut is *kamūn bīni*, literally "mushroom-like excrescence of the tamarisk".

Tamarisk played a major role in Sumerian and Akkadian magic and medicine, even entering cuneiform prose literature. According to the Babylonian dispute poem *The Tamarisk and the Date Palm* both trees grew in the garden of a king. One day they started a quarrel about which was more useful to man and god. After a lengthy exchange of words in which each tree praises its own virtues and points to the defects of the other, the date palm is judged the winner. The poem, dating to the beginning of the 2nd millennium BCE, shows how tamarisk was perceived: it gave shade, its wood served to manufacture divine statues and furniture, and it was used in purification ceremonies. The broom-like shoots of the tree were used to sprinkle water on the patient to cleanse him; cultic objects were decontaminated in much the same way.

Alum was the principal chemical used as a fixative (mordant) in dyeing wool and tanning leather, used as early as the 3rd millennium BCE. Its importance is evident, bearing in mind that garments and leather products were among the main export goods of ancient Mesopotamia. It is possible that the knowledge of the use of tamarisk gallnuts as a substitute for alum influenced how ancient Mesopotamians conceived purification. Like wool, which required boiling in a solution of alum to receive the dye, helping to brighten the ultimate colour, so man needed to be prepared, hence purified, to receive divine help.

Assyrian and Babylonian practitioners stated that tamarisk gallnuts were used for the same medical treatments as alum. These applications included cleaning the teeth before eating, potions for jaundice, obstruction of the bladder, bleeding from the anus and rectal inflammations, as well as potions for abnormal periods in women. An alum gargle was recommended to treat a disease affecting the mouth, nose and throat with bleeding caused by ulcers and inflammation of the gums (called in Akkadian *bu'šānu*). A salve with alum served to treat skin lesions and eye afflictions.

Twigs, bark, seeds, shoots, gallnuts, ash and manna of tamarisk were processed into a variety of medicaments. The finely crushed twigs were used to prepare salves for foot ulcers with scaly skin, poultices for gangrene of the feet and for problems with hearing.

"If a patient's right ear is hard-of-hearing, mix fresh tamarisk twigs with flour and place it in his ear." [1]

"If a skin lesion appears on the patient's feet causing fine fissures and cracks in the skin like eczema, the Akkadian name of the skin lesion is saġbānu *– if the lesion is full of watery fluid, the patient shall recover: crush tamarisk twigs, make a salve with cedar sap, apply it to the surface of the lesion."* [2]

"If a patient suffers from gangrene of the feet and the surface is black, crush tamarisk twigs to fine powder, sift the powder, mix it with flour and make a poultice with mustard water, apply it." [3]

Bark and gallnut were employed to treat afflictions of the skin. Assyrian and Babylonian practitioners wrote about the gallnut, which they called literally "fungus-like excrescence of tamarisk": *"Drug for skin rash – Boil it in water, sprinkle the water on the skin."* [4] The tree bark, possibly processed into a salve, was used to treat a reddish skin lesion that caused fever.

Other external application forms included a salve made of the seeds to reduce fever of the belly and an eye daub for watery eyes.

"Tamarisk seed – drug for fever of the belly whatever is its cause – to be crushed, to be anointed with sesame oil." [5]

"If a patient's right and left temples pulsate as if they were seized by a ghost, the ears ring and the eyes are full of tears, daub a mixture of copper filings and tamarisk on the patient's eyes." [6]

Manna and seed were wrapped in goat hair and carried around the neck to treat spider bites, or bring relief from the attack of *Lilû*, a demon associated with stormy winds in general, and the southwest wind in particular.

"Tamarisk manna – drug for a spider bite – to be wrapped in goat hair and hung around the patient's neck." [7]

"Tamarisk seed – drug for seizure of Lilû-demon – wrap it into fleece, hang it around the patient's neck." [8]

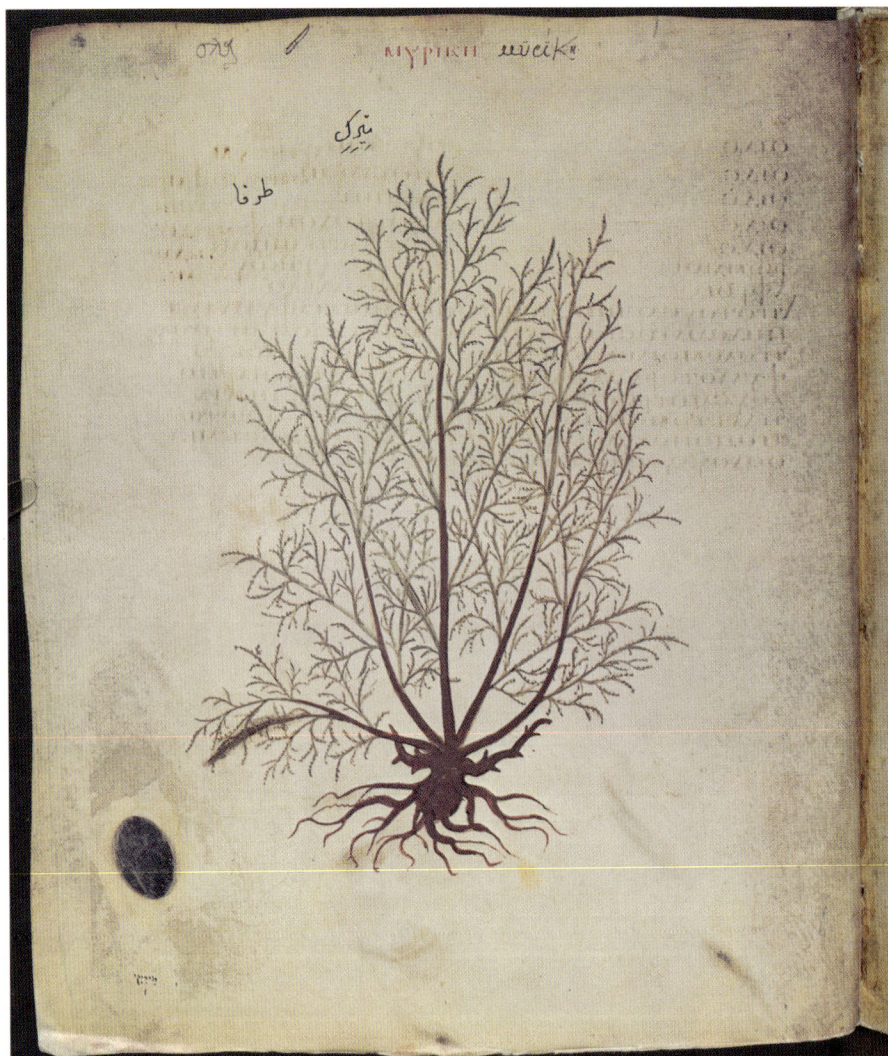

Tamarisk. Dioscorides Pedanius of Anazarbus, *De materia medica*. Constantinople, c. 512 CE. Cod. Med Gr. 1 (facs.), fol. 231v
© Österreichische Nationalbibliothek, Vienna

For internal applications, Assyrian and Babylonian practitioners used the seeds, twigs, shoots and gallnuts. The charcoal of tamarisk was processed into a wad of absorbent material for bleeding during pregnancy.

"If a woman suffers from bleeding during pregnancy, crush charred tamarisk, wrap it into fleece and introduce it into the vagina." [9]

A potion of the crushed seeds or twigs taken with beer or sesame oil was prescribed for the treatment of troubles with the bile, jaundice, constriction of the urethra, abnormal periods in women and anal disease with bleeding. The same potions served also as a tonic in case of fever, refusing food or loss of appetite.

"Tamarisk twigs – drug for the bile – to be crushed and taken with first-quality beer." [10]

"Tamarisk twigs – drug for jaundice – to be crushed and taken with beer." [11]

"If a patient's skin and face is yellow and he suffers from unexplained weight loss, this disease is called jaundice: crush tamarisk seed, have the patient drink it with beer or a mixture of sesame oil and beer." [12]

"If a man holds back something hard when urinating, crush tamarisk seed and have him drink it with beer." [13]

"If a woman suffers from abnormal blood flow which cannot be stopped, in order to stop it: crush tamarisk seed, have her drink it with beer on an empty stomach and the blood flow will stop." [14]

"Tamarisk seed – drug for anal disease with bleeding – to be ground up and taken with beer." [15]

For haemorrhoids, tamarisk products were processed into medium "acorn"-size suppositories and coated with suet mixed with various plant ingredients before application.

"If the patient suffers from anal troubles, viz. the haemorrhoids are bloated and full of blood, finely crush various plant ingredients, mix them with suet, and coat an acorn-sized suppository of tamarisk (gallnut) with the mixture, apply it repeatedly to his anus; then the patient will recover." [16]

Dioscorides (I, 87) describes the use of gallnuts for eye and mouth afflictions, to treat swellings, coughing up blood, bowel troubles, jaundice, spider bites, and women suffering from heavy periods. The bark served as mouthwash for toothaches, and the ash was applied as a pessary to control uterine discharges. Ibn al-Bayṭār (no. 1455) reports that tamarisk is suitable in haemorrhage and haemorrhoids, and is used for tooth decay, loose teeth, bile disorders and jaundice.

Tamarisk (*Tamarix aphylla*)
Courtesy: Ori Fragman-Sapir

BIBLIOGRAPHY AND NOTES

Chapter 1

Driver, G.R. 1944. Reginald Campbell Thompson. *Proceedings of the British Academy* 30: 447–85.

Smith, S. 1941. Dr. R. Campbell Thompson, F.B.A. *Nature* 3739: 799.

Thompson, R.C. 1924. *The Assyrian Herbal. A Monograph on the Assyrian Vegetable Drugs, The Subject Matter of which was Communicated in a Paper to the Royal Society, March 20, 1924.* London: Luzac & Co.

Thompson, R.C. 1949. *A Dictionary of Assyrian Botany.* London: British Academy.

Chapter 2

Asheri, D., Lloyd, A. & Corcella, A. 2007. *A Commentary on Herodotus Books I–IV.* Oxford: Oxford University Press.

Böck, B. 2014. *The Healing Goddess Gula. Towards an Understanding of Ancient Babylonian Medicine.* Leiden: Brill.

Civil, M. 1960. Prescriptions médicales sumériennes. *Revue d'assyriologie et d'archéologie orientale* 54: 57–75.

Dalley, S. 2013. *The Mystery of the Hanging Garden of Babylon.* Oxford: Oxford University Press.

Finkel, I. 2020. Assurbanipal's Library. An overview. In K. Ryholt & G. Barjamovic (eds), *Libraries before Alexandria: Ancient Near Eastern Traditions.* Oxford: Oxford University Press, 367–89.

Foster, K.P. 2004. The Hanging Gardens of Nineveh. *Iraq* 66: 207–20.

Fronzaroli, P. 1998. A pharmaceutical text at Ebla (TM.75.G.1623). *Zeitschrift für Assyriologie* 88: 225–39.

George, A.R. 2007. Babylonian and Assyrian: A history of Akkadian. In J.N. Postgate (ed.), *Languages of Iraq, Ancient and Modern.* London: British School of Archaeology in Iraq, 31–71.

Guest, E. & Al-Rawi, A. 1966. *Flora of Iraq* 1. Baghdad: Ministry of Agriculture & London: Royal Botanic Gardens, Kew.

Laursen, S. & Steinkeller, P. 2017. *Babylonia, the Gulf Region, and the Indus: Archaeological and Textual Evidence for Contact in the Third and Early Second Millennia B.C.* Winona Lake, IN: Eisenbrauns.

Liverani, M. 2013. *The Ancient Near East: History, Society and Economy.* London: Routledge.

Matthews, R. 2003. *Archaeology of Mesopotamia: Theories and Approaches.* London: Routledge.

Michalowski, P. 2004. Sumerian. In R.D. Woodard (ed.), *The Cambridge Encyclopedia of the*

World's Ancient Languages: The Ancient Languages of Mesopotamia, Egypt, and Aksum. Cambridge: Cambridge University Press, 19–59.

Oppenheim, A.L. 1977. *Ancient Mesopotamia: Portrait of a Dead Civilization.* Chicago, IL: University of Chicago Press.

Potts, D.T. 1997. *Mesopotamian Civilization: The Material Foundations.* Ithaca, NY: Cornell University Press.

Reade, J. 2000. Alexander the Great and the Hanging Gardens of Babylon. *Iraq* 62: 195–217.

Roaf, M. 1990. *Cultural Atlas of Mesopotamia and the Ancient Near East.* New York, NY: Andromeda Oxford.

[1] Köcher, F. 1964. *Babylonische Medizin in Texten und Untersuchungen* III. Berlin: de Gruyter, 248 iv: 34–8.

Chapter 3

Böck, B. 2020. Gedanken zu dem Drogen-Inventar aus Assur KADP 36 (VAT 8903). In S.M. Maul (ed.), *Assur-Forschungen* 2. Harrassowitz: Wiesbaden, 63–85.

Böck, B. 2021. Ancient Mesopotamian physic gardens: On medicinal plants, vegetables and spices. In H. Perdicoyianni-Paleologou (ed.), *Health, Disease, and Healing from Antiquity to Byzantium: Medicinal Foods, Plants and Spices* (Byzantinische Forschungen XXXIII). Amsterdam: Adolf M. Hakkert, 1–19.

Damerow, P. 2012. Sumerian beer: The origins of brewing technology in ancient Mesopotamia. *Cuneiform Digital Library Journal* 2012(2) (https://cdli.ucla.edu/pubs/cdlj/2012/cdlj2012_002.html).

Farber, W. & Freydank, H. 1977. Zwei medizinische Texte aus Assur. *Altorientalische Forschungen* 5: 255–8.

Geller, M.J. 2010. *Ancient Babylonian Medicine: Theory and Practice.* Chichester: Wiley.

Goltz, D. 1968. Mitteilungen über ein assyrisches Apothekeninventar. *Archives internationales d'histoire des sciences* 21: 95–114.

Köcher, F. 1978. Spätbabylonische Texte aus Uruk. In C. Habrich, F. Marguth & J.H. Wolf (eds), *Medizinische Diagnostik in Geschichte und Gegenwart: Festschrift für Heinz Goerke zum sechzigsten Geburtstag.* München: Werner Fritsch, 17–39.

Parpola, S. 1993. *Letters from Assyrian and Babylonian Scholars* (State Archives of Assyria 10). Helsinki: Helsinki University Press.

Sallaberger, W. 2012. Bierbrauen in Versen: Eine neue Edition und Interpretation der Ninkasi-Hymne. In C. Mittermayer & S. Ecklin (eds), *Altorientalische Studien zu Ehren von Pascal Attinger – mu-ni u_4 ul-li$_2$-a-aš ğa$_2$-ga$_2$-de$_3$* (Orbis Biblicus et Orientalis 256). Fribourg: Academic Press, 291–328.

Tavernier, J. 2008. KADP 36: Inventory, plant list, or lexical exercise. In R.D. Biggs, J. Myers & M.T. Roth (eds), *Proceedings of the 51st Rencontre Assyriologique Internationale Held at the Oriental Institute of the University of Chicago, July 18–22, 2005.* Chicago, IL: Oriental Institute of the University of Chicago, 191–202.

[1] Köcher, F. 1963b. *Babylonische Medizin in Texten und Untersuchungen* II. Berlin: de Gruyter, 171 l. 14', 22'. [2] Id. 1980b. *Babylonische Medizin in Texten und Untersuchungen* VI. Berlin: de Gruyter, 575 iii: 39.

Chapter 4

Böck, B. 2009. On medical technology in ancient Mesopotamia. In A. Attia & G. Buisson (eds), *Advances in Mesopotamian Medicine: From Hammurabi to Hippocrates.* Leiden: Brill, 105–28.

Böck, B. 2011. Sourcing, organising and administering medicinal ingredients. In K. Radner & E. Robson (eds), *The Oxford Handbook of Cuneiform Culture*. Oxford: Oxford University Press, 690–705.

Böck, B. 2015. Shaping texts and text genres: On the drug lore of Babylonian practitioners of medicine. *Aula Orientalis* 33: 21–37.

Bottéro, J. 1995. *Mesopotamian Culinary Texts* (Mesopotamian Civilizations 6). Winona Lake, IN: Eisenbrauns.

Damerow, P. 2012. Sumerian beer: The origins of brewing technology in ancient Mesopotamia. *Cuneiform Digital Library Journal* 2012(2) (https://cdli.ucla.edu/pubs/ cdlj/2012/cdlj2012_002.html).

Fincke, J.C. 2021. *An Ancient Mesopotamian Herbal Handbook. Volume 1: The Tablets*. Leuven: Peeters.

Fronzaroli, P. 1998. A pharmaceutical text at Ebla (TM.75.G.1623). *Zeitschrift für Assyriologie* 88: 225–39.

Goltz, D. 1974. *Studien zur altorientalischen und griechischen Heilkunde: Therapie, Arzneibereitung, Rezeptstruktur* (Sudhoffs Archiv. Beiheft 16). Wiesbaden: Steiner.

Herrero, P. 1984. *La thérapeutique mésopotamienne*. Paris: Editions Recherche sur les civilisations.

Ibn al-Bayṭār. 1877-1883. *Traité des simples par Ibn el-Beïtar*, 3 vols., translated by L. Leclerc, Paris: Imprimerie nationale.

Postgate, J.N. 1992. *Early Mesopotamia: Society and Economy at the Dawn of History*. London: Routledge.

Powell, M.A. 1993. Drugs and pharmaceuticals in Ancient Mesopotamia. In I. Jacob & W. Jacob (eds), *The Healing Past: Pharmaceuticals in the Biblical and Rabbinic World*. Leiden: Brill, 47–67.

Sallaberger, W. 2012. Bierbrauen in Versen: Eine neue Edition und Interpretation der Ninkasi-Hymne. In C. Mittermayer & S. Ecklin (eds), *Altorientalische Studien zu Ehren von Pascal Attinger – mu-ni u₄ ul-li₂-a-aš ǧa₂-ga₂-de₃* (Orbis Biblicus et Orientalis 256). Fribourg: Academic Press, 291–328.

Stadhouders, H. 2012. The pharmacopeial handbook Šammu šikinšu. *Journal des Médecines Cunéiformes* 19: 1–21.

Chapter 5
Language and comparing traditions

Aliotta, G., Piomelli, D., Pollio, A. & Touwaide, A. 2003. *Le piante medicinali del "Corpus hippocraticum"*. Naples: Guerini.

Bhayro, S., Haley, R., Kessel, G. & Pormann, P.E. 2013. The Syriac Galen palimpsest: Progress, prospects and problems. *Journal of Semitic Studies* 54: 131–48.

Fausti, D. 2004. Erbari illustrati su papiro e tradizione iconografica botanica. In I. Andorlini (ed.), *Testi medici su papiro: Atti del Seminario di Studio (Firenze, 3-4 giugno 2002)*. Florence: Istituto Papirologico G. Vitelli, 131–50.

Gutas, D. 1998. *Greek Thought, Arabic Culture: The Graeco-Arabic Translation Movement in Baghdad and Early 'Abbāsid Society (2nd–4th / 8th–10th centuries)*. London: Routledge.

Hardy, G. & Totelin, L. 2016. *Ancient Botany*. London: Routledge.

Jones-Lewis, M. 2016. Pharmacy. In G.L. Irby (ed.), *A Companion to Science, Technology, and Medicine in Ancient Greece and Rome*. New York, NY: Wiley, 402–17.

Kessel, G. 2018. Syriac medicine. In D. King (ed.), *The Syriac World*. London: Routledge, 438–59.

Levey, M. 1966. *The Medical Formulary or Aqrābādhīn of al-Kindī*. Madison, WI: University of Wisconsin Press.

Levey, M. 1973. *Early Arabic Pharmacology*. Leiden: Brill.

Meyerhof, M. 1931. Alî at-Tabarî's "Paradise of Wisdom", one of the oldest Arabic compendiums of medicine. *Isis* 16: 6–54.

Powell, M.A. 2003–2005. Obst und Gemüse (Fruits and vegetables). A.I. Mesopotamien. In *Reallexikon der Assyriologie und Vorderasiatischen Archäologie* 10. Berlin: de Gruyter, 13–22.

Riddle, J.M. 1985. *Dioscorides on Pharmacy and Medicine*. Austin, TX: University of Texas Press.

Saad, B. & Said, O. 2011. *Greco-Arab and Islamic Herbal Medicine: Traditional System, Ethics, Safety, Efficacy, and Regulatory Issues*. Hoboken, NJ: John Wiley.

Touwaide, A. 2019. Botany. In S. Lazaris (ed.), *A Companion to Byzantine Science*. Leiden: Brill, 302–53.

Ullmann, M. 1978. *Islamic Medicine*. Edinburgh: Edinburgh University Press.

Plants

Al-Rawi, A. 1964. *Wild Plants of Iraq*. Baghdad: Ministry of Agriculture.

Christenhusz, M.J.M., Fay, M.F. & Chase, M.W. 2017. *Plants of the World: An Illustrated Encyclopaedia of Vascular Plants*. London: Royal Botanic Gardens, Kew.

Farjon, A. & Filer, D. 2013. *An Atlas of the World's Conifers: An Analysis of Their Distribution, Biogeography, Diversity and Conservation Status*. Leiden: Brill.

Flora of Iraq

    Bor, N.L. 1968. *Flora of Iraq* 9. Baghdad: Ministry of Agriculture & London: Royal Botanic Gardens, Kew.

    Ghazanfar, S.A. & Edmondson, J.R. 2013. *Flora of Iraq* 5(2). Baghdad: Ministry of Agriculture & London: Royal Botanic Gardens, Kew.

    Ghazanfar, S.A. & Edmondson, J.R. 2016. *Flora of Iraq* 5(1). Baghdad: Ministry of Agriculture & London: Royal Botanic Gardens, Kew.

    Ghazanfar, S.A., Edmondson, J.R. & Hind, D.J.N. 2019. *Flora of Iraq* 6. Baghdad: Ministry of Agriculture & London: Royal Botanic Gardens, Kew.

    Guest, E. & Al-Rawi, A. 1966. *Flora of Iraq* 1. Baghdad: Ministry of Agriculture & London: Royal Botanic Gardens, Kew.

    Guest, E. & Al-Rawi, A. 1966. *Flora of Iraq* 2. Baghdad: Ministry of Agriculture & London: Royal Botanic Gardens, Kew.

    Townsend, C.C. & Guest, E. 1974. *Flora of Iraq* 3. Baghdad: Ministry of Agriculture & London: Royal Botanic Gardens, Kew.

    Townsend, C.C., Guest, E. & Omar, S.A. 1980. *Flora of Iraq* 4(1) & 4(2). Baghdad: Ministry of Agriculture & London: Royal Botanic Gardens, Kew.

    Townsend, C.C., Guest, E., Omar, S.A. & Khayat, A.H. 1985. *Flora of Iraq* 8. Baghdad: Ministry of Agriculture & London: Royal Botanic Gardens, Kew.

Ghorbani, A., Wieringa, J.J., Boer, H.J.D., Porck, H., Kardinaal, A. & Andel, T.V. 2018. Botanical and floristic composition of the historical herbarium of Leonhard Rauwolf collected in the Near East (1573–1575). *Taxon* 67: 565–80.

Quattrocchi, U. 2017. *CRC World Dictionary of Plant Names: Common Names, Scientific Names, Eponyms, Synonyms, and Etymology*. New York: Routledge.

Walter, T., Ghorbani, A. & van Andel, T. 2022. The emperor's herbarium: The German physician Leonhard Rauwolf (1535?–96) and his botanical field studies in the Middle East. *History of Science* 60(1), 130–151.

Archaeobotany

ADEMNES: The GRF (German Research Foundation) and University of Tübingen funded Archaeobotanical Database of Eastern Mediterranean and Near Eastern Sites, last update 2015. www.ademnes.de.

Charles, M.P. 1990. Agriculture in Lowland Mesopotamia in the Late Uruk–Early Dynastic Period. PhD Thesis, University of London.

Charles, M.P. 2023. Plant remains from the 6H House. In J.N. Postgate, *Abu Salabikh Excavations*, Vol. 5. London: British School of Archaeology in Iraq, 394-405.

Curtis, R.I. 2001. *Ancient Food Technology*. Leiden: Brill.

Ellison, R., Renfrew, J., Brothwell, D. & Seeley, N. 1978. Some food offerings from Ur, excavated by Sir Leonard Woolley, and previously unpublished. *Journal of Archaeological Science* 5: 167–77.

Helbaek, H. 1965. Early Hassunan vegetable remains at es-Sawwan near Samarra. *Sumer* 20: 45–8.

Helbaek, H. 1966. The plant remains from Nimrud. In M.E.L. Mallowan (ed.), *Nimrud and Its Remains*. Edinburgh: Edinburgh University Press, 613–20.

Helbaek, H. 1972. Samarran irrigation agriculture at Chogha Mami in Iraq. *Iraq* 34: 35–48.

Neef, R. 1989. Plant remains from archaeological sites in lowland Iraq: Hellenistic and Neobabylonian Larsa. In J.-L. Huot (ed.), *Larsa, travaux de 1985*. Paris: Editions Recherche sur les civilisations, 151–61.

Neef, R. 1991. Plant remains from archaeological sites in lowland Iraq: Tell el 'Oueli. In J.-L. Huot (ed.), *'Oueili, Travaux de 1985*. Paris: Editions Recherche sur les civilisations, 321–9.

Nesbitt, M. 1995. Plants and people in ancient Anatolia. *Biblical Archaeologist* 58: 68–81.

Nesbitt, M. 2003–2005. Obst und Gemüse (Fruits and vegetables). B. Archäobotanisch. In *Reallexikon der Assyriologie und Vorderasiatischen Archäologie* 10. Berlin: de Gruyter, 26–30.

Nicholson, P.T. & Shaw, I. (eds), 2000. *Ancient Egyptian Materials and Technology*. Cambridge: Cambridge University Press.

Peyronel, L., Vacca, A. & Wachter-Sarkady, C. 2014. Food and drink preparation at Ebla, Syria. New data from the Royal Palace G (c. 2450–2300 BC). *Food & History* 14: 3–38.

Prance, G. & Nesbitt, M. 2004. *The Cultural History of Plants*. New York: Routledge.

Rivera, D., Matilla, G., Obón, C. & Alcaraz, F. 2012. *Plants and Humans in the Near East and the Caucasus. Ancient and Traditional Uses of Plants as Food and Medicine. An Ethnobotanical Diachronic Review*. Murcia: Universidad de Murcia.

Savard, M. 2005. Epipalaeolithic to Early Neolithic Subsistence Strategies in the Northern Fertile Crescent: The Archaeobotanical Remains from Hallan Çemi, Demirköy, M'lefaat and Qermez Dere. PhD Thesis, University of Cambridge.

Savard, M., Nesbitt, M. & Gale, R. 2003. Archaeobotanical evidence for Early Neolithic diet and subsistence at M'lefaat (Iraq). *Paléorient* 29: 93–106.

van Zeist, W. 1991. Lower Mesopotamian plant remains from the fourth century B.C. and the first/second century A.D. In L. de Meyer (ed.), *Northern Akkad Project Reports 5*, Mesopotamian History and Environment, Series 1. Ghent: University of Ghent, 61–9.

van Zeist, W. & Vynckier, J. 1984. Palaeobotanical investigations of Tell ed-Der. In L. de Meyer (ed.), *Tell ed-Der IV*. Leuven: Peeters, 119–43.

Zohary, D. & Spiegel-Roy, P. 1975. Beginnings of fruit growing in the Old World. *Science* 187: 319–27.

Zohary, D., Hopf, M. & Weiss, E. 2012. *Domestication of Plants in the Old World: The Origin and Spread of Domesticated Plants in South-West Asia, Europe, and the Mediterranean Basin*. Oxford: Oxford University Press.

Traditional uses

Al-Douri, N.A. 2000. Survey of medicinal plants and their traditional uses in Iraq. *Pharmaceutical Biology* 38: 74–9.

Al-Douri, N.A. 2014. Some important medicinal plants in Iraq. *ASJ International Journal of Advances in Herbal and Alternative Medicine* 2: 10–20.

Al-Rawi, A. & Chakravarty, H.L. 1964. *Medicinal Plants of Iraq. Agriculture Technical Bulletin* 15. Baghdad: Government Press.

Dafni, A. & Böck, B. 2019. Medicinal plants of the Bible – Revisited. *Journal of Ethnobiology and Ethnomedicine* 15 (57). https://doi.org/10.1186/s13002-019-0338-8.

Dauncey, E.A. & Howes, M.-J.R. 2020. *Plants that Cure: Plants as a Source of Medicines – from Pharmaceuticals to Herbal Remedies.* Kew: Royal Botanic Gardens, Kew.

Ghazanfar, S.A. 1994. *Handbook of Arabian Medicinal Plants.* Boca Raton, FL: CRC Press.

Ghazanfar, S.A. & Thorogood, C. (eds), 2022. *A Herbal of Iraq.* Kew: Royal Botanic Gardens, Kew.

Rivera, D., Matilla, G., Obón, C. & Alcaraz, F. 2012. *Plants and Humans in the Near East and the Caucasus. Ancient and Traditional Uses of Plants as Food and Medicine. An Ethnobotanical Diachronic Review.* Murcia: Universidad de Murcia.

[1] Gurney, O.R. & Finkelstein, J.J. 1957. *The Sultantepe Tablets.* London: British Institute of Archaeology at Ankara, 93, 63'–64'. [2] Ibid., 92 i: 12–15. [3] Scurlock, J.A. 2014. *Sourcebook for Ancient Mesopotamian Medicine.* Atlanta, GA: SBL Press, 319 i: 19. [4] Bottéro, J. 1995. *Mesopotamian Culinary Texts* (Mesopotamian Civilizations 6). Winona Lake, IN: Eisenbrauns, 42–3.

Herbal (in addition to bibliography for Chapters 1–5)
General

Al-Kindī. 1966. *The Medical Formulary or Aqrābādhīn of al-Kindī,* translated by M. Levey. Madison, WI: University of Wisconsin Press.

Beekes, R. 2010. *Etymological Dictionary of Greek,* Leiden: Brill.

Brown, F., Driver, S.R. & Briggs, C. 1906. *Hebrew and English Lexicon of the Old Testament.* Oxford: Clarendon Press.

Dioscorides. 2017. *Pedanius Dioscorides of Anazarbus: De materia medica,* translated by L.Y. Beck. Hildesheim: Olms–Weidmann.

Hort, A. 1916. *Theophrastus: Enquiry Into Plants.* Cambridge MA: Harvard University Press & London: William Heinemann.

Ibn al-Bayṭār. 1877–1883. *Traité des simples par Ibn el-Beïtar,* 3 vols., translated by L. Leclerc. Paris: Imprimerie nationale.

Lane, E.W. 1863–1893. *Arabic–English Lexicon.* London & Edinburgh: Williams and Norgate.

Lev, E. & Amar, Z. 2008. *Practical Materia Medica of the Medieval Eastern Mediterranean According to the Cairo Genizah.* Leiden: Brill.

Löw, I. 1924–1934. *Die Flora der Juden.* Wien: Löwit.

Onions, C.T. 1966. *Oxford Dictionary of English Etymology.* Oxford: Oxford University Press.

Quattrocchi, U. 2017. *CRC – World Dictionary of Plant Names.* Boca Raton: Routledge.

Sauerhoff, F. 2004. *Etymologisches Wörterbuch der Pflanzennamen.* Stuttgart: Wissenschaftliche Verlagsgesellschaft mbH.

Soden, W. von. 1959–1981. *Akkadisches Handwörterbuch.* Wiesbaden: Harrassowitz.

Sokoloff, M. 2009. *A Syriac Lexicon.* Winona Lake, IN: Eisenbrauns & Piscataway, NJ: Gorgias Press.

Vaan, M. de. 2008. *Etymological Dictionary of Latin and the Other Italic Languages.* Leiden: Brill.

Wehr, H. 1993. *A Dictionary of Modern Written Arabic.* Ithaca, NY: Spoken Languages Services.

Principal cuneiform sources

Gurney, O.R. & Finkelstein, J.J. 1957. *The Sultantepe Tablets.* London: British Institute of Archaeology at Ankara.

Köcher, F. 1955. *Keilschrifttexte zur assyrische-babylonischen Drogen- und Pflanzenkunde*. Berlin: Akademie-Verlag.

Köcher, F. 1963a. *Babylonische Medizin in Texten und Untersuchungen* I. Berlin: de Gruyter.

Köcher, F. 1963b. *Babylonische Medizin in Texten und Untersuchungen* II. Berlin: de Gruyter.

Köcher, F. 1964. *Babylonische Medizin in Texten und Untersuchungen* III. Berlin: de Gruyter.

Köcher, F. 1971. *Babylonische Medizin in Texten und Untersuchungen* IV. Berlin: de Gruyter.

Köcher, F. 1980a. *Babylonische Medizin in Texten und Untersuchungen* V. Berlin: de Gruyter.

Köcher, F. 1980b. *Babylonische Medizin in Texten und Untersuchungen* VI. Berlin: de Gruyter.

Thompson, R.C. 1923. *Assyrian Medical Texts*. London: Oxford University Press.

Thompson, R.C. 1949. *A Dictionary of Assyrian Botany*. London: British Academy.

Black bryony

Pliny the Elder. 1951. *Natural History Vol. VI, Books XX–XXIII*, translated by W.H.S. Jones. Cambridge, MA: Harvard University Press.

Renner, S.S., Scarborough J., Schaefer, H., Paris, H.S. & Janick, J. 2008. Dioscorides's bruonia melaina is *Bryonia alba*, not *Tamus communis*, and an illustration labelled bruonia melaina in the *Codex Vindobonensis* is *Humulus lupulus* not *Bryonia dioica*. In M. Pitrat (ed.), *Cucurbitaceae 2008*. Avignon: INRA, 273–80.

[1] Gurney & Finkelstein 1957, 93 ll. 63'–64'. [2] Thompson 1923, 52,5 obv. 4', 12'. [3] Weiher, E. von. 1988. *Spätbabylonische Texte aus Uruk III*. Berlin: Gebr. Mann Verlag, 106 obv. 11. [4] Gurney & Finkelstein 1957, 92 iii: 19'. [5] Köcher 1980a, 421 i: 22'. [6] Id. 1963a, 1 ii: 51. [7] Id. 1980b, 574 ii: 2, 3. [8] Ibid., 578 ii: 11, 65. [9] Ibid., 578 iii: 7, 15. [10] Thompson 1923, 85,1 ii: d-e. [11] Köcher 1955, 1 v: 28. [12] Id. 1971, 379 ii: 46 (with duplicates). [13] Id. 1964, 237 i: 21', 37'.

Black nightshade

Böck, B. 2021. Mind-altering plants in Babylonian medical sources. In D. Stein, S.K. Costello & K.P. Foster (eds), *The Routledge Companion to Ecstatic Experience in the Ancient World*. London: Routledge, 121–37.

[1] Gurney & Finkelstein 1957, 93 obv. 19–20. [2] Köcher 1980b, 515 i: 63'–64'. [3] Id. 1963b, 159 i: 21-23, 27. [4] Id. 1980b, 578 iv: 17, 23. [5] Id. 1963a, 1 ii: 49. [6] Thompson 1923, 62,1 iii: 3'. [7] Geller, M.J. 2005. *Renal and Rectal Disease Texts*. Berlin: de Gruyter, 44 l. 16, 46 l. 31. [8] Köcher 1971, 396 ii: 25', iii: 10–11. [9] Id. 1963b, 146 obv. 15'–16'. [10] Ibid., 3 ii: 36, 41. [11] Id. 1971, 380 rev. 35; ibid., 381 iii: 29-30. [12] Id. 1963a, 1 ii: 11. [13] Id. 1964, 248 iv: 13, 21, 15. [14] Steinert, U. 2012. K. 263+10934. *Sudhoffs Archiv* 96: 65 l. 1, 16. [15] Gurney & Finkelstein 1957, 92 i: 14.

Camel thorn

Alhagi graecorum https://powo.science.kew.org/taxon/urn:lsid:ipni.org:names:473468-1

Bodenheimer, F.S. 1947. The manna of Sinai. *Biblical Archaeologist* 10: 2–6.

Civil, M. 1987a. Feeding Dumuzi's sheep: The lexicon as a source of literary inspiration. In F. Rochberg-Halton (ed.), *Language, Literature, and History: Philological and Historical Studies Presented to Erica Reiner*. New Haven, CT: American Oriental Society, 37–55.

Civil, M. 1987b. Ur III Bureaucracy: Quantitative aspects. In M. Gibson & R.D. Biggs (eds), *The Organization of Power: Aspects of Bureaucracy in the Ancient Near East* (Studies in Ancient Oriental Civilization 46). Chicago, IL: University of Chicago Press, 43–53.

Donkin, R.A. 1980. Mannas of Western and Central Asia and of North Africa. In R.A. Donkin (ed.) *Manna: An Historical Geography*. Biogeographica 17. Dordrecht: Springer, 12–73.

Meyerhof, M. 1947. The earliest mention of a manniparous insect. *Isis* 37: 32–6.

Ramezany, F., Kiyani, N. & Khademizadeh, M. 2013. Persian manna in the past and the present: An overview. *American Journal of Pharmacological Sciences* 1: 35–7.

Soriga, E. 2017. A diachronic view on fulling technology in the Mediterranean and the ancient Near East: Tools, raw materials and natural resources for the finishing of textiles. In S. Gaspa, C. Michel & M.-L. Nosch (eds), *Textile Terminologies from the Orient to the Mediterranean and Europe, 1000 BC to 1000 AD*. Lincoln, NE: Zea Books, 24–46.

Tavassoli, A.P., Anushiravani, M., Hoseini, S.M., Nikakhtar, Z., Baghdar, H.N., Ramezani, M., Ayati, Z., Amiri, M.S., Sahebkar, A. & Emami, S.A. 2020. Phytochemistry and therapeutic effects of *Alhagi* spp. and tarangabin in traditional and modern medicine: A review. *Journal of Herbmed Pharmacology* 9: 86–104.

Verma, R.L. 1990. Al-Biruni on the science of medicine. In W.H. Abdi (ed.), *Interaction Between Indian and Central Asian Science and Technology in Medieval Times*, Vol. 2. New Delhi: Indian National Science Academy, 86–96.

[1] Köcher 1955, 6 vi: 14'. [2] Ibid., 6 vi: 16'. [3] Ibid., 6 vi: 15'. [4] Id. 1963a, 1 i: 24. [5] Ibid., 1 i: 41. [6] Scurlock, J.A. 2014. *Sourcebook for Ancient Mesopotamian Medicine*. Atlanta, GA: SBL Press, 413 l. 49, 415 l. 84. [7] Köcher 1980a, 432 i: 18'. [8] Id. 1963a, 1 i: 10. [9] Ibid., 11 l. 23, 27. [10] Thompson 1923, 20,1 i: 24', 28'–29'. [11] Id. 1964, 260 l. 1, 4.

Cedar

Al-Rawi, F.N.H. & George, A.R. 2014. Back to the cedar forest: The beginning and end of Tablet V of the standard Babylonian epic of Gilgameš. *Journal of Cuneiform Studies* 66: 69–90.

Böck, B. 2021. Mind-altering plants in Babylonian medical sources. In D. Stein, S.K. Costello & K.P. Foster (eds), *The Routledge Companion to Ecstatic Experience in the Ancient World*. London: Routledge, 121–37.

George, A.R. 2003. *The Epic of Gilgamesh*. London: Penguin.

Heimpel, W. The sun at night and the doors of heaven in Babylonian texts. *Journal of Cuneiform Studies* 38: 127–51.

Horowitz, W. 1998. *Mesopotamian Cosmic Geography*. Winona Lake, IN: Eisenbrauns.

Scurlock, J.A. 2006. *Magico-Medical Means of Treating Ghost-Induced Illnesses in Ancient Mesopotamia*. Leiden: Brill.

Starr, I. 1983. *The Rituals of the Diviner* (Bibliotheca Mesopotamica 12). Malibu, CA: Undena.

[1] BM 38583 obv. 3'. [2] Köcher 1963a, 3 ii: 12, 13. [3] Thompson 1923, 52,5 obv. 4'–5', 12'. [4] BM 59593 l. 39. [5] Köcher 1964, 237 iv: 15, 22. [6] Scurlock, J.A. 2006, 446. [7] Scurlock, J.A. 2014. *Sourcebook for Ancient Mesopotamian Medicine*. Atlanta, GA: SBL Press, 383 ii: 72'.

Colocynth

Breslin, C.A.Y. 1986. Abu Hanifah al-Dinawari's Book of Plants: An Annotated English Translation of the Extant Alphabetical Portion. MA Thesis, University of Arizona.

Chomicki, C., Schaefer, G. & Renner, S.S. 2019. Origin and domestication of Cucurbitaceae crops: Insights from phylogenies, genomics and archaeology. *New Phytologist* 226: 1240–55.

Ebeling, E. 1949. Ein Rezept zum Würzen von Fleisch. *Orientalia NS* 18: 171–2.

Ghazanfar, S.A. & A.M.A. Al-Sabahi. 1993. Medicinal plants of northern and central Oman (Arabia). *Economic Botany* 47: 89–98.

Janick, J., Paris, H.S. & Parrish, D.C. 2016. The cucurbits of Mediterranean antiquity: Identification of taxa from ancient images and descriptions. *Vegetation History and Archaeobotany* 25: 405–14.

Paris, H.S., Daunay, M. & Janick, J. 2012. Occidental diffusion of cucumber (*Cucumis sativus*) 500–1300 CE: Two routes to Europe. *Annals of Botany* 109: 117–26.

Rosner, F. 1995. *Maimonides, Medical Writings, the Glossary of Drug Names*. Translated and annotated from Max Meyerhof's French edition. Haifa: Maimonides Research Institute.

Stol, M. 1987. The Cucurbitaceae in the cuneiform texts. *Bulletin on Sumerian Agriculture* 3: 81–92.

[1] Gurney & Finkelstein 1957, 93, 66'–67'. [2] Köcher 1963b, 3 ii: 12, 152 i: 9'. [3] Ibid., 124 ii: 19, 21. [4] Ibid., 124 ii: 19, 26. [5] Thompson 1923, 74 iii: 13, 15. [6] Ibid., 75, iv: 17, 18. [7] BM 103386 rev. 36–7. [8] Geller, M.J. 2005. *Renal and Rectal Disease Texts*. Berlin: de Gruyter, 44 l. 18. [9] Köcher 1971, 396 iii: 6. [10] Id. 1980b, 578 i: 30, 36. [11] Id. 1963b, 159 ii: 12–16, 19. [12] Köcher 1980b, 575 i: 1, 12–13. [13] Ibid., 575 ii: 26. [14] Ibid., 578 i: 45. [15] Id. 1963b., 159 ii: 49–50, iii: 7. [16] BM 51246+53217 rev. 13'–14'.

Coriander

Bottéro, J. 1995. *Mesopotamian Culinary Texts* (Mesopotamian Civilizations 6). Winona Lake, IN: Eisenbrauns.

Potts, D.T. 1997. *Mesopotamian Civilization: The Material Foundations*. Ithaca, NY: Cornell University Press.

[1] Köcher 1963b, 159 ii: 25, 36. [2] Id. 1963a, 1 iii: 38.

Cumin

Bottéro, J. 1995. *Mesopotamian Culinary Texts* (Mesopotamian Civilizations 6). Winona Lake, IN: Eisenbrauns.

Catagnoti, A. 2010. Il lessico dei vegetali ad Ebla 3. Piante aromatiche (parte I): Cumino e timo. *Quaderni del Dipartimento di Linguistica – Università di Firenze* 20: 143–9.

Pliny the Elder. 1950. *Natural History Vol. V, Books XVII–XIX*, translated by H. Rackham. Cambridge, MA: Harvard University Press.

Potts, D.T. 1997. *Mesopotamian Civilization: The Material Foundations*. Ithaca, NY: Cornell University Press.

[1] Köcher 1963a, 1 i: 48. [2] Ibid., 1 ii: 20. [3] Scheil, V. 1918. Notules. *Revue d'assyriologie et d'archéologie* 15: 76 obv. 14', 16'. [4] Köcher 1963b, 146 obv. 8', 17'. [5] Labat, R. 1957. Remèdes assyriens contre les affections de l'oreille, d'après un inédit du Louvre (AO. 6774). *Rivista degli studi orientali* 32: 113–114 ii: 1–4.

Date palm

Flowers, J.M., Hazzouri, K.M., Gros-Balthazard, M., Mo, Z., Koutroumpa, K., Perrakis, A., Ferrand, S., Khierallah, H.S., Fuller, D.Q., Aberlenc, F. & Fournaraki, C., 2019. Cross-species hybridization and the origin of North African date palms. *Proceedings of the National Academy of Sciences* 116: 1651–8.

Gros-Balthazard, M., Galimberti, M., Kousathanas, A., Newton, C., Ivorra, S., Paradis, L., Vigouroux, Y., Carter, R., Tengberg, M., Battesti, V. & Santoni, S. 2017. The discovery of wild date palms in Oman reveals a complex domestication history involving centers in the Middle East and Africa. *Current Biology* 27: 2211–18.

Herodotus. 1920. *The Histories* (Loeb Classical Library, Volume I). Harvard, MA: Harvard
University Press, 193.

Jennings, R.P., Shipton, C., Al-Omari, A., Alsharekh, A.M., Crassard, R., Groucutt, H.
& Petraglia, M.D. 2013. Rock art landscapes beside the Jubbah palaeolake, Saudi
Arabia. *Antiquity* 87: 666–83.

Landsberger, B. 1967. *The Date Palm and Its By-products According to the Cuneiform Sources* (Archiv
für Orientforschung Beiheft 17). Graz: Weidner.

Moore Jr, H.E. & Dransfield, J. 1979. The typification of Linnaean palms. *Taxon* 28: 59–70.

Potts, D.T. 1997. *Mesopotamian Civilization: The Material Foundations*. Ithaca, NY: Cornell
University Press.

Roth, M.T. 1997. *Law Collections from Mesopotamia and Asia Minor*. Atlanta, GA: Scholars Press.

Streck, M.P. 2004. Dattelpalme und Tamariske in Mesopotamian nach dem akkadischen
Streitgespräch. *Zeitschrift für Assyriologie* 94: 250–90.

[1] Köcher 1980a, 506 obv. 10'. [2] Thompson 1923, 75 iii: 26, 27. [3] Köcher 1963b, 124 ii: 19, 30.
[4] Id. 1963a, 3 ii: 12; id. 1963b, 152 i: 11'. [5] Id. 1980b, 578 iv: 17, 22. [6] Ibid., 510 i: 21–2. [7] Id.
1963a, 99, 19–21. [8] Id. 1964, 237 i: 21, 25. [9] Id. 1971, 379 iii: 33'. [10] Id. 1963b, 159 i: 21, 25–6.

False carob

Civil, M. 1987a. Feeding Dumuzi's sheep: The lexicon as a source of literary inspiration. In F.
Rochberg-Halton (ed.), *Language, Literature, and History: Philological and Historical Studies
Presented to Erica Reiner*. New Haven, CT: American Oriental Society, 37–55.

Civil, M. 1987b. Ur III Bureaucracy: Quantitative aspects. In McG. Gibson & R.D. Biggs (eds),
The Organization of Power: Aspects of Bureaucracy in the Ancient Near East (Studies in Ancient
Oriental Civilization 46). Chicago, IL: University of Chicago Press, 43–53.

Maekawa, K. 1990. Cultivation methods in the Ur III period. *Bulletin on Sumerian Agriculture* 5:
115–45.

Scurlock, J.A. 2006. *Magico-Medical Means of Treating Ghost-Induced Illnesses in Ancient
Mesopotamia*. Leiden: Brill.

Soriga, E. 2017. A diachronic view on fulling technology in the Mediterranean and the ancient
Near East: Tools, raw materials and natural resources for the finishing of textiles.
In S. Gaspa, C. Michel & M.-L. Nosch (eds), *Textile Terminologies from the Orient to the
Mediterranean and Europe, 1000 BC to 1000 AD*. Lincoln, NE: Zea, 24–46.

[1] Worthington, M. 2005. Edition of UGU (=*BAM* 480 etc.). *Journal des Médecines Cunéiformes* 5:
20 l. 155', 167'–168'. [2] Köcher 1963b, 124 i: 33, 37. [3] Id. 1963a, 1 i: 42. [4] Id. 1980a, 423 i: 19'. [5]
Geller, M.J. 2005. *Renal and Rectal Disease Texts*. Berlin: de Gruyter, 44 l. 18, 24. [6] Steinert, U.
2012. K. 263+10934. *Sudhoffs Archiv* 96: 65: 1, 11. [7] Köcher 1963a, 1 i: 7. [8] Scurlock, J.A. 2006, 443
ll. 1-4. [9] George, A.R. 2016, *Mesopotamian Incantations and Related Texts in the Schøyen Collection*
(Cornell University Studies in Assyriology and Sumerology 32). Bethesda, MD: CDL Press, 56.

Fig

Bottéro, J. 2004. *The Oldest Cuisine in the World: Cooking in Mesopotamia*. Chicago, IL: University
of Chicago Press.

Flaishman, M.A., Rodov, V. & Stover, E. 2008. The fig: botany, horticulture, and breeding.
Horticultural Reviews 34: 113–97.

Postgate, J.N. 1987. Notes on fruit in the cuneiform sources. *Bulletin on Sumerian Agriculture* 3:
115–44.

Stol, M. 2004. Wirtschaft und Gesellschaft in altbabylonischer Zeit. In P. Attinger, W. Sallaberger & M. Wäfler (eds), *Mesopotamian: Die altbabylonischer Zeit* (Orbis Biblicus et Orientalis 160/4). Fribourg: Academic Press, 643–975.

[1] Köcher 1980a, 480 ii: 12, 15.

Flax, Linseed

Linum usitatissimum https://powo.science.kew.org/taxon/urn:lsid:ipni.org:names:544772-1
Postgate, J.N. 1992. *Early Mesopotamia: Society and Economy at the Dawn of History.* London: Routledge.
Potts, D.T. 1997. *Mesopotamian Civilization: The Material Foundations.* Ithaca, NY: Cornell University Press.
Quillien, L. 2014. Flax and linen in the first millennium Babylonia BC. In M. Harlow, C. Michel & M. Nosche (eds), *Prehistoric, Ancient Near Eastern and Aegean Textiles and Dress: An Interdisciplinary Anthology.* Oxford: Oxbow, 271–96.
Schweinfurth, G. 1884. Further discoveries in the flora of ancient Egypt. *Nature* 29: 312–15.

[1] Geller, M.J. 2005. *Renal and Rectal Disease Texts.* Berlin: de Gruyter, 222 ll. 14'-15'. [2] Böck, B. 2010. 2.10.9 Schlaganfall. In B. Janowski & D. Schwemer (eds), *Texte zur Heilkunde: Texte aus der Umwelt des Alten Testaments NF 5.* Gütersloh: Gütersloher Verlagshaus, 95 2.10.9 ll. 11'-12'.

Garden rocket

Eruca vesicaria https://apps.kew.org/herbcat/detailsQuery.do?barcode=K001039942
Böck, B. 2021. Mind-altering plants in Babylonian medical sources. In D. Stein, S.K. Costello & K.P. Foster (eds), *The Routledge Companion to Ecstatic Experience in the Ancient World.* London: Routledge, 121–37.
Bottéro, J. 1995. *Mesopotamian Culinary Texts* (Mesopotamian Civilizations 6). Winona Lake, IN: Eisenbrauns.
Pliny the Elder. 1951. *Natural History Vol. VI, Books XX–XXIII,* translated by W.H.S. Jones. Cambridge, MA: Harvard University Press.
Powell, O. 2003. *Galen: On the Properties of Foodstuffs.* Cambridge: Cambridge University Press.

[1] Biggs, R.D. 1967. *ŠÀ.ZI.GA Ancient Mesopotamian Potency Incantations.* Locust Valley, NY: J.J. Augustin, 52 text AMT 88,3:1–8. [2] Köcher 1971, 380 rev. 47. [3] Ibid., 380 rev. 51. [4] Heeßel, N.P. 2008. Warzen, Beule und Narben. Eine Sammlung medizinischer rezepte und physiognomischer Beobachtungen aus Assur. In R.J. van der Spek (ed.), *Studies in Ancient Near Eastern World View and Society.* Bethesda, MD: CDL Press, 167 ll. 5–7. [5] Geller, M.J. & Panayotov, S.V. 2020. *Mesopotamian Eye Disease Texts.* Berlin: de Gruyter, 68 l. 40'.

Garlic

Block, E. 2010. *Garlic and Other Alliums: The Lore and the Science.* London: Royal Society of Chemistry.
Böck, B. 2021. Ancient Mesopotamian physic gardens: On medicinal plants, vegetables and spices. In H. Perdicoyianni-Paleologou (ed.), *Health, Disease, and Healing from Antiquity to Byzantium: Medicinal Foods, Plants and Spices* (Byzantinische Forschungen XXXIII). Amsterdam: Adolf M. Hakkert, 1–19.
Germer, R. 2008. *Handbuch der altägyptischen Heilpflanzen.* Wiesbaden: Harrassowitz.
Postgate, J.N. 1987. Some vegetables in the Assyrian sources. *Bulletin on Sumerian Agriculture* 3: 93–100.

Stol, M. 1987. Garlic, onion, leek. *Bulletin on Sumerian Agriculture* 3: 57–80.

Waetzoldt, W. 1987. Knoblauch und Zwiebeln nach den Texten des 3. Jt. *Bulletin on Sumerian Agriculture* 3: 23–56.

[1] Köcher 1963a, 1 i: 54. [2] Id. 1963b, 158 iii: 16'. [3] Labat, R. 1957. Remèdes assyriens contre les affections de l'oreille, d'après un inédit du Louvre (AO. 6774). *Rivista degli studi orientali* 32: 116 iv: 10'. [4] Köcher 1983b, 575 ii: 26. [5] Id. 1963a, 1 i: 28. [6] Id. 1983b, 578 ii: 67, 70. [7] Id. 1963b, 159 ii: 25, 41. [8] Thompson 1923, 85,1 ii: 1-2.

Greater plantain, Ribwort

Böck, B. 2014. *The Healing Goddess Gula: Towards an Understanding of Ancient Babylonian Medicine*. Leiden: Brill.

Böck, B. in press. How to understand the mode of action Babylonian medicines take: The identification of Akkadian *lišān kalbi* "Dog's tongue" with *Cynoglossum* and *Plantago*. In J. Quack (ed.), *Wirksamkeit von Heilbehandlungen in den Medizinalsystemen des Alten Orients und Ägyptens*.

[1] Böck 2014: 142–4. [2] Ibid., 152. [3] Ibid., 148–51. [4] Ibid., 145–7. [5] Ibid., 144–5. [6] Ibid., 151. [7] Ibid., 147–8. [8] Ibid., 156.

Henbane

Böck, B. 2021. Mind-altering plants in Babylonian medical sources. In D. Stein, S.K. Costello & K.P. Foster (eds), *The Routledge Companion to Ecstatic Experience in the Ancient World*. London: Routledge, 121–37.

Sibbing-Plantholt, I. 2014. A new look at the Kassite medical letters, and an edition of Šumu-libši Letter N 969. *Zeitschrift für Assyriologie* 104: 171–81.

[1] Köcher 1955, 1 v: 1. [2] Gurney & Finkelstein 1957, 93 obv. 42. [3] Köcher 1963b, 124 i: 17, 24. [4] Id. 1955, 1 v: 7. [5] Ibid., 1 v: 8. [6] Ibid., 1 v: 6. [7] Gurney & Finkelstein 1957, 93 obv. 35'-36'.

Juniper

Heimpel, W. 2011. Twenty-eight trees growing in Sumer. In D.I. Owen (ed.), *Garšana Studies*. Bethesda, MD: CDL Press, 75–152.

Jacobsen, T. 1987. *The Harps that Once: Sumerian Poetry in Translation*. New Haven, CT: Yale University Press, 183–204.

Jones, J.F. 1857. *Memoirs of Baghdad, Kurdistan and Turkish Arabia*. Selections from the Records of the Bombay Government, no. XLIII. Bombay: Education Society's Press.

Postgate, J.N. 2014–2016. Wacholder. *Reallexikon der Assyriologie Assyriologie und Vorderasiatischen Archäologie* 14. Berlin: de Gruyter, 606–7.

[1] Gurney & Finkelstein 1957, 92 ii: 3. [2] BM 38583 obv. 12'–13'. [3] Köcher 1980b, 578 iii: 7, 11. [4] Ibid., 578 ii: 67, 68. [5] Ibid., 575 i: 37, 44. [6] Ibid., 574 ii: 10, 12. [7] Ibid., 575 ii: 26, 29. [8] Id. 1964, 240 l. 26', 27'. [9] Scurlock, J.A. 2014. *Sourcebook for Ancient Mesopotamian Medicine*. Atlanta, GA: SBL Press, 474 l. 53, 54. [10] Gurney & Finkelstein 1957, 94 obv. 13'. [11] Köcher 1980a, 503 iii: 79', iv: 5. [12] Thompson 1923, 75 iv: 17. [13] Köcher 1963a, 1 ii: 7. [14] Ibid., 3 iii: 47–48. [15] Ibid., 1 i: 16.

Leek

Böck, B. 2021. Ancient Mesopotamian physic gardens: On medicinal plants, vegetables and spices. In H. Perdicoyianni-Paleologou (ed.), *Health, Disease, and Healing from Antiquity to Byzantium: Medicinal Foods, Plants and Spices* (Byzantinische Forschungen XXXIII). Amsterdam: Adolf M. Hakkert, 1–19.

Bottéro, J. 1995. *Mesopotamian Culinary Texts* (Mesopotamian Civilizations 6). Winona Lake, IN: Eisenbrauns.

Postgate, J.N. 1987. Some vegetables in the Assyrian sources. *Bulletin on Sumerian Agriculture* 3: 93–100.

Stol, M. 1987. Garlic, onion, leek. *Bulletin on Sumerian Agriculture* 3: 57–80.

[1] Schwemer, D. 2013. Prescriptions and rituals for happiness, success, and divine favor: The compilation A 522 (BAM 318). *Journal of Cuneiform Studies* 65: 188 iii: 22. [2] Köcher 1963a, 1 iii: 38. [3] Id. 1980b, 575 i: 1, 19. [4] Scurlock, J.A. 2014. *Sourcebook for Ancient Mesopotamian Medicine*. Atlanta, GA: SBL Press, 496 ii: 35. [5] Köcher 1980b, 543 ii: 24, 34. [6] Id. 1964, 240 rev. 59–60', 64'.

Liquorice

Civil, M. 1960. Prescriptions médicales sumériennes. *Revue d'assyriologie et d'archéologie orientale* 54: 57–75.

Ghazanfar, S.A. 1994. *Handbook of Arabian Medicinal Plants*. Boca Raton, FL: CRC. Hort, A. 1916. *Theophrastus: Enquiry into Plants*. Books 6–9. Cambridge, MA: Harvard University Press.

Schopen, A. 1983. *Traditionelle Heilmittel in Jemen*. Wiesbaden: Franz Steiner.

[1] Köcher 1971, 393 obv. 4–7. [2] Gurney & Finkelstein 1957, 92 ii: 4. [3] Köcher 1980b, 578 i: 23. [4] Gurney & Finkelstein 1957, 92 ii: 13. [5] Köcher 1980b, 574 i: 4, 8. [6] Id. 1963b, 159 ii: 12, 17. [7] Ibid., 124 i: 33, 38. [8] Ibid., 124 ii: 36, 41. [9] Id. 1980b, 579 i: 1. [10] Wasserman, N. 2010. From the notebook of a professional exorcist. In D. Shehata, F. Weierhäuse & K.V. Zand (eds), *Von Göttern und Menschen: Beiträge zu Literatur und Geschichte des Alten Orients. Festschrift für Brigitte Groneberg*. Leiden: Brill, 331.

Onion

Böck, B. 2021. Ancient Mesopotamian physic gardens: On medicinal plants, vegetables and spices. In H. Perdicoyianni-Paleologou (ed.), *Health, Disease, and Healing from Antiquity to Byzantium: Medicinal Foods, Plants and Spices* (Byzantinische Forschungen XXXIII). Amsterdam: Adolf M. Hakkert, 1–19.

Bottéro, J. 1995. *Mesopotamian Culinary Texts* (Mesopotamian Civilizations 6). Winona Lake, IN: Eisenbrauns.

Brewster, J.L. 2008. *Onions and Other Vegetable Alliums*. Wallingford: CABI.

Postgate, J.N. 1987. Some vegetables in the Assyrian sources. *Bulletin on Sumerian Agriculture* 3: 93–100.

Stol, M. 1987. Garlic, onion, leek. *Bulletin on Sumerian Agriculture* 3: 57–80.

[1] Geller, M.J. & Panayotov, S.V. 2020. *Mesopotamian Eye Disease Texts*. Berlin: de Gruyter, 62 l. 26'.

Pomegranate

Chandra, R., Babu, D.K., Jadhav, V.T., Jaime, A. & Silva, T.D. 2010. Origin, history and domestication of pomegranate. *Fruit, Vegetable and Cereal Science and Biotechnology* 2: 1–6.
Heimpel, W. 2011. Twenty-eight trees growing in Sumer. In D.I. Owen (ed.), *Garšana Studies*. Bethesda, MD: CDL Press, 75–152.
Levin, G.M. 2006. *Pomegranate*. Tempe, AZ: Third Millennium Publications.
Stol, M. 1980–1083. Leder (industrie). *Reallexikon der Assyriologie Assyriologie und Vorderasiatischen Archäologie* 6. Berlin: de Gruyter, 527–43.
Van de Mieroop, M. 1992. Wood in the Old Babylonian texts from southern Babylonia. *Bulletin on Sumerian Agriculture* 6: 155–61.
Ward, C., 2003. Pomegranates in eastern Mediterranean contexts during the Late Bronze Age. *World Archaeology* 34: 529–41.

[1] Köcher 1980b, 578 iv: 5. [2] Hunger, H. 1976. *Spätbabylonische Texte aus Uruk I*. Berlin: Gebrüder Mann, 52 l. 29, 31. [3] BM 38583 rev. 3 obv. 12', rev. 3. [4] Hunger, H. 1976. *Spätbabylonische Texte aus Uruk I*. Berlin: Gebrüder Mann, 54 l. 29, 31. [5] Ibid., 54 l. 35, 38. [6] Labat, R. 1957. Remèdes assyriens contre les affections de l'oreille, d'après un inédit du Louvre (AO. 6774). *Rivista degli studi orientali* 32: 114, ii: 8'–10'. [7] Köcher 1980a, 503 iii: 12, 16. [8] Scurlock, J.A. 2014. *Sourcebook for Ancient Mesopotamian Medicine*. Atlanta, GA: SBL Press, 381 i: 40'. [9] Ibid., 437 l. 63–69. [10] Thompson 1923, 74 iii: 13, 16. [11] Köcher 1963b, 124 ii: 19, 33. [12] Ibid., 124 ii: 36, 42. [13] Ibid., 124 ii: 47–48. [14] Id. 1980b, 575 ii: 50–51. [15] Id. 1963b, 159 ii: 25, 28–31. [16] Id. 1964, 237 iv: 15, 18. [17] Biggs, R.D. 1967. *ŠÀ.ZI.GA Ancient Mesopotamian Potency Incantations*. Locust Valley, NY: J.J. Augustin, 70 text KAR 61 ll. 1–6, 9–10.

Poplar

Civil, M. 1960. Prescriptions médicales sumériennes. *Revue d'assyriologie et d'archéologie orientale* 54: 57–75.
Foster, B.R. 2005. *Before the Muses: An Anthology of Akkadian Literature*. Bethesda, MD: CDL Press.
Heimpel, W. 2011. Twenty-eight trees growing in Sumer. In D.I. Owen (ed.), *Garšana Studies*. Bethesda, MD: CDL Press, 75–152.
Postgate, J.N. 2003–2005. Pappel. *Reallexikon der Assyriologie Assyriologie und Vorderasiatischen Archäologie* 10. Berlin: de Gruyter, 329.
Powell, M.A. 1992. Timber production in Presargonic Lagaš. *Bulletin on Sumerian Agriculture* 6: 99–122.
Qasem, J.R.S. 2020. *The Coloured Atlas of Medicinal and Aromatic Plants of Jordan and their Uses*. Newcastle upon Tyne: Cambridge Scholars, 283.
Volk, K. 1995. *Inanna und Shukaletuda: Zur historisch-politischen Deutung eines sumerischen Literaturwerkes*. Wiesbaden: Harrassowitz.

[1] Köcher 1964, 237 iv: 15, 21. [2] Id. 1963b, 159 ii: 49, iii: 8. [3] Köcher 1963a, 3 i: 48. [4] Id. 1980a, 494 ii: 41, 42. [5] Thompson 1923, 74 ii: 34, iii: 1. [6] Geller, M.J. & Panayotov, S.V. 2020. *Mesopotamian Eye Disease Texts*. Berlin: de Gruyter, 67-68 l. 40', 44'. [7] Köcher 1980b, 543 ii: 24, 31. [8] Ibid., 549 i: 10'–12'. [9] Ibid., 575 i: 37, 47. [10] Ibid., 575 i: 25, 31. [11] Id. 1980a, 480 ii: 67. [12] Id. 1963b, 146 obv. 15'; id. 1953, *Keilschrifturkunden aus Boghazköi* 37, Berlin: Akademie-Verlag, 2 obv. 16'. [13] Geller, M.J. 2005. *Renal and Rectal Disease Texts*. Berlin: de Gruyter, 44 l. 18, 23. [14] Köcher 1980a, 503 iii: 12, 13.

Red bryony

Kahl, O. 2021. *Sābūr Ibn Sahl. Dispensatorium Parvum (Al-Aqrābādhīn al-Saghīr)* (Islamic Philosophy, Theology and Science. Texts and Studies 16). Leiden: Brill.

Renner, S.S., Scarborough J., Schaefer, H., Paris, H.S. & Janick, J. 2008. Dioscorides's bruonia melaina is *Bryonia alba*, not *Tamus communis*, and an illustration labelled bruonia melaina in the *Codex Vindobonensis* is *Humulus lupulus* not *Bryonia dioica*. In M. Pitrat (ed.), *Cucurbitaceae 2008*. Avignon: INRA, 273–80.

[1] Gurney & Finkelstein 1957, 93 rev. 58'–59'. [2] Köcher 1980b, 578 ii: 67, 69. [3] Ibid., 578 ii: 23, 60. [4] Ibid., 578 iii: 7, 23 (restored). [5] Ibid., 574 i: 42. [6] Id. 1963a, 92 ii': 22', 38'. [7] Id. 1964, 316 iii: 17'–18'. [8] Id. 1963b, 159, ii: 43–5. [9] Geller, M.J. 2005. *Renal and Rectal Disease Texts*. Berlin: de Gruyter, 44 l. 18, 30. [10] Weiher, E. von. 1988. *Spätbabylonische Texte aus Uruk III*. Berlin: Gebr. Mann Verlag, 106 obv. 9. [11] Köcher 1955, 1 v: 26-27. [12] Gurney & Finkelstein 1957, 92 i: 15. [13] Köcher 1963a, 33 ll. 1–2, 8. [14] Id. 1980a, 421 i: 21' (with duplicates). [15] Id. 1964, 237 i: 21', 37'. [16] Id. 1971, 381 iii: 20. [17] Abusch, T. & Schwemer, D. 2011. *Corpus of Mesopotamian Anti-Witchcraft Rituals, Vol. I*. Leiden: Brill, 187 ll. 17'–30'. [18] Id., 191 l. 24. [19] Köcher 1955, 1 v: 27.

Sesame

Abusch, T. 2016. *The Magical Ceremony Maqlû*. Leiden: Brill.

Alster, B. 1997. *Proverbs from Ancient Sumer, Vol. I*. Bethesda, MD: CDL Press.

Bedigian, D. 2004. History and lore of sesame in southwest Asia. *Economic Botany* 58: 329–53.

Bedigian, D. & Harlan, J.R. 1986. Evidence for cultivation of sesame in the ancient world. *Economic Botany* 40: 137–54.

Maul, S.M. & Strauß, R. 2011. *Ritualbeschreibungen und Gebete I* (Keilschriftliche Texte aus Assur literarischen Inhalts 4). Wiesbaden: Harrassowitz, 61.

Postgate, J.N. 1992. *Early Mesopotamia: Society and Economy at the Dawn of History*. London: Routledge.

Potts, D.T. 1997. *Mesopotamian Civilization: The Material Foundations*. Ithaca, NY: Cornell University Press.

Stol, M. 1985. Remarks on the cultivation of sesame and the extraction of its oil. *Bulletin on Sumerian Agriculture* 2: 119–26.

[1] Geller, M.J. & Panayotov, S.V. 2020. *Mesopotamian Eye Disease Texts*. Berlin: de Gruyter, 58 ll. 12–13. [2] Ibid., 62 l. 26'. [3] Ibid., 170 l. 81', 83'. [4] Scurlock, J.A. 2014. *Sourcebook for Ancient Mesopotamian Medicine*. Atlanta, GA: SBL Press, 447, 451 ii: 25. [5] Ibid., 318 i: 1–4, 310 i: 19.

Sweet flag

Avadhani, M., Selvaraj, C.I., Rajasekharan, P.E. & Tharachand, C. 2013. The sweetness and bitterness of sweet flag (*Acorus calamus* L.) – A review. *Research Journal of Pharmaceutical, Biological and Chemical Sciences* 4: 598–610.

Jursa, M. 2009. Die Kralle des Meeres und andere Aromata. In W. Arnold, M. Jursa, W.W. Müller & S. Procházka (eds), *Philologisches und Historisches zwischen Anatolien und Sokotra. Analecta Semitica In Memoriam Alexander Sima*. Wiesbaden: Harrassowitz, 147–80.

Middeke-Conlin, R. 2014. The scents of Larsa: A study of the aromatics industry in an Old Babylonian kingdom. *Cuneiform Digital Library Journal* 2014(1): 1–53.

Pliny the Elder. 1945. *Natural History Vol. IV, Books XII–XVI*, translated by H. Rackham. Cambridge, MA: Harvard University Press.

Waetzoldt, H. 1992. 'Rohr' und dessen Verwendungsweisen anhand der neusumerischen Texte aus Umma. *Bulletin on Sumerian Agriculture* 6: 125–46.

[1] Köcher 1980a, 503 i: 20', 25'. [2] Ibid., 423 i: 15'. [3] Ibid., 423 i: 17'. [4] Gurney & Finkelstein 1957, 94 rev. 48'. [5] Scurlock, J.A. 2014. *Sourcebook for Ancient Mesopotamian Medicine*. Atlanta, GA: SBL Press, 596 ll. 52-53.

Tamarisk

Ahmad, M., Zafar, M. & Sultana, S. 2009. *Salvadora persica, Tamarix aphylla* and *Zizyphus mauritiana*: Three woody plant species mentioned in Holy Quran and Ahadith and their ethnobotanical uses in north western part (DI Khan) of Pakistan. *Pakistan Journal of Nutrition* 8: 542–7.

Baum, B.B. 1978. *The Genus Tamarix*. Jerusalem: Israel Academy of Sciences and Humanities.

Böck, B. 2018. Zur Weitergabe und Verbreitung altmesopotamischen medizinischen Wissens: Die Verwendung von Alaun und Schmirgel in diachronischer Perspektive. In K. Kleber, G. Neumann & S. Paulus (eds), *Grenzüberschreitungen: Studien zur Kulturgeschichte des Alten Orients. Festschrift für Hans Neumann anlässlich seines 65. Geburtstages am 9. Mai 2018*. Münster: Zaphon, 59–77.

Bodenheimer, F.S. 1947. The manna of Sinai. *Biblical Archaeologist* 10: 2–6.

Danin, A. 1972. A sweet exudate of *Hammada*: Another source of manna in Sinai. *Economic Botany* 26: 373–5.

Donkin R.A. 1980. Mannas of Western and Central Asia and of North Africa. In R.A. Donkin (ed.) *Manna: An Historical Geography*. Biogeographica 17. Dordrecht: Springer, 12–73.

Heimpel, W. 2011. Twenty-eight trees growing in Sumer. In D.I. Owen (ed.), *Garšana Studies*. Bethesda, MD: CDL Press, 75–152.

Powell, M.A. 1992. Timber production in Presargonic Lagaš. *Bulletin on Sumerian Agriculture* 6: 99–122.

Soriga, E. 2017. A diachronic view on fulling technology in the Mediterranean and the ancient Near East: Tools, raw materials and natural resources for the finishing of textiles. In S. Gaspa, C. Michel & M.-L. Nosch (eds), *Textile Terminologies from the Orient to the Mediterranean and Europe, 1000 BC to 1000 AD*. Lincoln, NE: Zea, 24–46.

Streck, M.P. 2004. Dattelpalme und Tamariske in Mesopotamian nach dem akkadischen Streitgespräch. *Zeitschrift für Assyriologie* 94: 250–90.

[1] Labat, R. 1957. Remèdes assyriens contre les affections de l'oreille, d'après un inédit du Louvre (AO. 6774). *Rivista degli studi orientali* 32: 116, iv: 10', 12'. [2] Thompson 1923, 74 iii: 13, 75 iii: 20. [3] Köcher 1963b, 124 i: 33, 34. [4] Id. 1971, 379 iii: 47'. [5] Id. 1963a, 1 iii: 29. [6] Thompson 1923, 14,5: 6-7. [7] Köcher 1980a, 423 i: 24'; id. 1963a, 1 i: 39. [8] Gurney & Finkelstein 1957, 92 i: 19; Thompson, R.C. 1902. *Cuneiform Texts from Babylonian Tablets in the British Museum XIV*. London: Harrison and Sons 1902, 23 K.9283 l. 19. [9] Köcher 1964, 237 iv: 15, 20; id. 1980a 427 l. 12'. [10] Id. 1963a, 1 i: 34; Scheil, V. 1916. Un document médical assyrien, *Revue d'assyriologie* 13: 37 rev. 22'. [11] Scheil, ibid., 37 rev. 29'. [12] Köcher 1980b, 578, iii: 7, 14, 15. [13] Geller, M.J. 2005. *Renal and Rectal Disease Texts*. Berlin: de Gruyter, 44 l. 18, 20. [14] Steinert, U. 2012. K. 263+10934. *Sudhoffs Archiv* 96: 65: 11, 18. [15] Köcher 1963a, 1 iii: 4; 7 iv: 46'. [16] Heeßel, N.P. 2020. A new medical therapeutic text on rectal disease. In S.V. Panayotov & L. Vacín (eds), *Mesopotamian Medicine and Magic: Studies in Honor of Markham J. Geller*. Leiden & Boston: Brill, 320, 326 obv. ll. 36'–38'.

IMAGE SOURCES

Front cover: Relief panel from the Northwest Palace at Nimrud (ca. 883–859 BCE) showing
 a winged supernatural figure kneeling before a stylized tree ©The Metropolitan
 Museum of Art, New York; Gift of John D. Rockefeller Jr., 1932 (The Collection of
 Ancient Near Eastern Art, Accession Number: 32.143.14)

31.172.1, © The Metropolitan Museum of Art, New York

54.117.7, © The Metropolitan Museum of Art, New York

B17710 (U.10935*bis*), © University of Pennsylvania Museum of Archaeology and Anthropology

BM 124799, © The British Museum

BM 124821, © The British Museum

BM 124821, © The British Museum

BM 124825,a © The British Museum

BM 124939,a © The British Museum

BM 46226, © The British Museum

BM 63869, © The British Museum

Cod. Med Gr. 1 (facs.), fol. 58v, fol. 91v, fol. 190v, fol. 231v, fol. 292v, Photo: © Österreichische
 Nationalbibliothek, Vienna

Drawing of a seal impression showing a banquet scene with a woman and a man drinking
 from a vessel (after P. Amiet, La glyptique mésopotamienne archaïque, Paris 1980, n°
 1171); courtesy of Ignacio Márquez Rowe

K. 4345+, © The British Museum

K. 67+, © The British British Museum

M.652, fol. 49r, fol. 80r, fol. 251v, © The Morgan Library & Museum

Map of Mesopotamia, © Oxford University Press, reproduced with the permission of the
 Licensor through PLSclear, map produced by Oxford Cartographers 98800

Map of Mesopotamia – Overview, © Oxford University Press, reproduced with the permission
 of the Licensor through PLSclear, map produced by Oxford Cartographers 98800

MM 501, courtesy of the Museu de Montserrat, Barcelona, Photo: Barbara Böck

MS Or. 289, fol. 22r, fol. 34r, fol. 45r, fol. 45v, fol. 85r, fol. 92r, fol. 92v, fol. 125r, © Leiden,
 Universiteitsbibliotheek, CC BY 4.0

Reconstruction of racks for drying herbs, © Barbara Böck

VA Ass 4346 and VA Ass 4347, © Staatliche Museen zu Berlin, Vorderasiatisches Museum,
 Photo: Olaf M. Teßmer

VA Bab 4108.001, VA Bab 7584, VA Bab 7587, VA Bab 7570, VA Bab 7562, VA Bab 7562, VA Bab
 7594, © Staatliche Museen zu Berlin, Vorderasiatisches Museum, Photo: Olaf M. Teßmer

VAT 8246, © Staatliche Museen zu Berlin, Vorderasiatisches Museum, Photo: Olaf M. Teßmer

VAT 8903, © Staatliche Museen zu Berlin, Vorderasiatisches Museum, Photo: Olaf M. Teßmer

ACKNOWLEDGEMENTS

We are very grateful to Elizabeth Dauncey, Henry Oakeley, Nicholas Postgate and Ignacio Márquez Rowe for their careful reading of a draft text; we are responsible for any remaining errors.

We thank the Ministerio de Ciencia , Innovación y Universidades (MCIU) and the Ministerio de Ciencia e Innovación (MICIN), Spain for supporting our research and the publication of this book through the grants PGC2018-097821-B-I00 (MCIU/AEI/10.13039/50110, Feder, UE – Una manera de hacer Europa) and PID2021-125678NB-I00 (MCIN/AEI/10.13039/501100011033, Feder, UE – Una manera de hacer Europa).

INDEX TO DISEASES

GENERAL INDEX